T0399828

Sliding Mode Control of Semi-Markovian Jump Systems

Sliding Mode Control of Semi-Markovian Jump Systems

Baoping Jiang and Hamid Reza Karimi

CRC Press
Taylor & Francis Group
Boca Raton London New York

CRC Press is an imprint of the
Taylor & Francis Group, an **informa** business

First edition published 2022
by CRC Press
6000 Broken Sound Parkway NW, Suite 300, Boca Raton, FL 33487-2742

and by CRC Press
2 Park Square, Milton Park, Abingdon, Oxon, OX14 4RN

ISBN: 978-0-367-56503-9 (hbk)
ISBN: 978-0-367-56504-6 (pbk)
ISBN: 978-1-003-09804-1 (ebk)

Typeset in Times
by codeMantra

Contents

Preface

Recently, semi-Markovian jump systems (S-MJSs) have received wide attention due to their feasibility in modeling practical systems, for instance, electrical systems, economics and mechanics, that suffer from abrupt structural changes caused by phenomena such as random failures or repairs, changes in subsystem interconnections and so on. Nowadays, the analysis and synthesis of S-MJSs is becoming more and more rich, and many important results have been achieved. However, due to the generality that the transition rates in S-MJSs are time-varying and hard to obtain in practice, the fundamental issues of stochastic stability and stabilization of such systems are still challenging. Hence, we make an attempt to tackle the analysis and synthesis of a class of continuous-time S-MJSs with generally uncertain transition rates.

Sliding mode control has unique advantages in dealing with nonlinear complex systems: the sliding mode control model has the properties such as fast response and good transient performance, strong robustness to system perturbations and uncertainties, which has attracted great attention in the control community since its appearance. For sliding mode control of stochastic systems, the two-step design become more complicated than normal state-space systems; it is necessary to consider the jumping effect caused by the switching modes, the finite-time reachability due to jumping rules, the controller design with deficiency-mode transition rates information, etc. This monograph contains valuable references and knowledge to help the relevant researchers to explore these issues and carry out further research in the area.

The purposes of this monograph are to present latest development and literature review on sliding mode control of S-MJSs, which involves problems such as stochastic stability analysis, fuzzy integral sliding mode control, finite-time sliding mode control, adaptive sliding mode control and decentralized sliding mode control for S-MJSs, and their applications in robotic manipulator and circuit systems, for instance. The contents are also suitable for a one-semester graduate course.

More specially, in this monograph, basic concepts and results on stochastic stability of S-MJSs are first presented in Chapter 1. In Chapter 2, based on the former fundamental results, the problem of stochastic stability for S-MJSs with generally uncertain transition rates (TRs) is investigated through linear matrix inequality (LMI) technique. Chapter 3 deals with the issue of robust fuzzy integral sliding mode control for continuous-time Takagi–Sugeno fuzzy model-based S-MJSs, and the assumption that input matrices are the same with full column rank is removed. Chapter 4 treats finite-time sliding mode control of continuous-time S-MJs with immeasurable premise variables via fuzzy approach, and LMI conditions are proposed to guarantee boundedness performance both at the reaching phase and at the sliding motion phase. In Chapter 5, observer-based adaptive sliding mode control for nonlinear Takagi–Sugeno fuzzy model-based S-MJSs with immeasurable premise variables is investigated, LMI conditions for stochastic stability with an H_∞ performance disturbance attenuation level γ of the sliding mode dynamics and error dynamics are developed, and an adaptive controller is synthesized to ensure finite-time reachability of a predefined sliding surface. Chapter 6 presents a decentralized adaptive sliding mode control scheme for the stabilization of large-scale semi-Markovian jump-inter-connected systems, in which adaptive law is designed to compensate dead-zone input nonlinearity and unknown interconnections. Chapter 7 gives reduced-order sliding mode control approach to stabilize delayed-switching S-MJSs, LMI conditions are proposed for mean-square exponential stability analysis, and adaptive controller is designed for finite-time reachability purpose.

MATLAB® is a registered trademark of The MathWorks, Inc. For product information,
    please contact:
    The MathWorks, Inc.
    3 Apple Hill Drive
    Natick, MA 01760-2098 USA
    Tel: 508-647-7000
    Fax: 508-647-7001
    E-mail: info@mathworks.com
    Web: www.mathworks.com
    Baoping Jiang
    Hamid Reza Karimi

Acknowledgment

This work is supported in part by The National Natural Science Foundation of China under Grant 62003231; partially supported by The Natural Science Foundation of Jiangsu Province under Grant BK20200989; Partially funded by the NCF for colleges and universities in Jiangsu Province under Grant 20KJB120005 and the China Postdoctoral Science Foundation (2021M692369), and partially supported by the Italian Ministry of Education, University and Research through the Project "Department of Excellence LIS4.0-Lightweight and Smart Structures for Industry 4.0".

Authors

Dr. Baoping Jiang
School of Electronic and Information Engineering, Suzhou University of Science and Technology, Suzhou 215009, China (e-mails: baopingj@163.com)
Baoping Jiang received the Ph.D. degree in control theory from the Ocean University of China, Qingdao, China, in 2019. From 2017 to 2019, he was a joint training Ph.D. candidate in the Department of Mechanical Engineering, Politecnico di Milano, Milan, Italy. In 2019, he joined the Suzhou University of Science and Technology, Suzhou, China, where he is an associate professor. His research interests include sliding mode control, stochastic systems, etc.

Dr. Hamid Reza Karimi
Department of Mechanical Engineering, Politecnico di Milano, 20156 Milan, Italy, Email: hamidreza.karimi@ polimi.it
Hamid Reza Karimi received the B.Sc. (First Hons.) degree in power systems from the Sharif University of Technology, Tehran, Iran, in 1998, and the M.Sc. and Ph.D. (First Hons.) degrees in control systems engineering from the University of Tehran, Tehran, in 2001 and 2005, respectively. He is currently Professor of Applied Mechanics in the Department of Mechanical Engineering, Politecnico di Milano, Milan, Italy. His current research interests include control systems and mechatronics with applications to automotive control systems, robotics, vibration systems and wind energy.

Dr. Karimi is currently serving as the Chief Editor, Technical Editor and Associate Editor for some international journals. He is a Fellow of The International Society for Condition Monitoring (ISCM), a member of The Agder Academy of Science and Letters and also a member of the IEEE Technical Committee on Systems with Uncertainty, the Committee on Industrial Cyber-Physical Systems, the IFAC Technical Committee on Mechatronic Systems, the Committee on Robust Control and the Committee on Automotive Control. Dr. Karimi has been awarded as the 2016–2020 Web of Science Highly Cited Researcher in Engineering, the 2020 IEEE Transactions on Circuits and Systems Guillemin-Cauer Best Paper Award, August-Wilhelm-Scheer Visiting Professorship Award, JSPS (Japan Society for the Promotion of Science) Research Award, and Alexander-von-Humboldt-Stiftung Research Award, for instance. He has also participated as General Chair, keynote/plenary speaker, distinguished speaker or program chair for many international conferences in the areas of control systems, robotics and mechatronics.

1 Introduction

1.1 SLIDING MODE CONTROL

Sliding mode control (SMC) has received considerable attention since its first appearance in the 1950s by Emelyanov and then was developed by Utkin, which has been proven to be an effective robust control strategy for incompletely modeled or nonlinear systems [1–3]. Essentially, SMC is a special kind of nonlinear control, and its nonlinearity lies in control of discontinuity, which makes SMC differ from other control methods – that is, the structure of the controlled system remains unchanged in the dynamic process, but can constantly change according to the current state of the system (such as its deviation and derivative, etc.) to meet a desired requirement. The consequence is that the system-state trajectories will be forced onto a specified manifold and kept a satisfactory sliding motion. Since the sliding mode can be designed independent of the object parameters and disturbances, the SMC has the advantages such as fast response, insensitivity to parameter changes and disturbances, without the need of on-line identification for the system, and simple physical implementation. However, SMC has one main unavoidable defect: the sliding motion is difficult to strictly remain to the equilibrium point after the state trajectory reaches onto the sliding manifold, but to cross back and forth on both sides of the sliding surface, which is the resource of chattering problem. Despite the disadvantage, the researches and applications of SMC are rich, and the developments are continuing.

1.1.1 BASIC CONCEPTS OF SMC

The continue-time case:
Variable structure control with sliding mode is called sliding mode variable structure control or SMC. The system structure changes constantly according to switching signals on both sides of $S(x)=0$ in the state space. The principle of switching is the called control strategy, which guarantees the existence of sliding mode dynamics. Accordingly, $S = S(x)$ and $S(x)=0$ are called switching functions and switching surfaces, respectively. The sliding mode means the movement that the motion point (state variable) of the system is attracted to the region as it approaches this region. The motion of the system in the sliding mode region is called "sliding mode motion," which has the property that the motion has nothing to do with the plant changed parameters and disturbances.

For a general nonlinear system of the form

$$\dot{x} = f(x,u,t), x \in \mathbb{R}^n, u \in \mathbb{R}^m, t \in \mathbb{R}, \tag{1.1}$$

where $x(t)$ is the system-state vector, and $u(t)$ is the control input.

Now, given a switching function of the form $S(x) = [s_1(x) \ s_2(x) \ \cdots \ s_m(x)]^T$, we need to design a sliding mode controller $u(t) = [u_1(t) \ u_2(t) \ \cdots \ u_m(t)]^T$ that

$$u_i(t) = \begin{cases} u_i^+(t), \ s_i(t) > 0, \\ u_i^-(t), \ s_i(t) < 0, \end{cases}$$

in which $u_i^+ \neq u_i^-$, such that the following three conditions are satisfied:

1. The sliding mode motion exists, that is, $S(x)\dot{S}(x) \leq 0$;
2. The reaching condition is satisfied; that is, all system trajectories outside the sliding surface $S(x) = 0$ will reach onto the predefined sliding surface in finite time subsequently maintained;
3. The dynamics in sliding surface $S(x) = 0$, that is, the sliding mode dynamics, is stable with some specified performances.

The discrete-time case:

Discrete SMC is also known as SMC of discrete-time systems; that is, it extends the SMC theory in continuous-time systems to discrete-time systems in order to meet the requirements in current digital computer control systems. The study of discrete SMC began in the 1980s [4], and its control strategies can be generally divided into *inequality arrival conditions* and *equality arrival conditions*.

In the discrete-time case, the possibility that the system-state trajectories strictly slide on the switching surface is almost zero. Therefore, the SMC of discrete-time system comes up with the issue of "quasi-sliding mode"; that is for the discrete-time systems, the SMC cannot produce an ideal sliding motion, but can only produce the quasi-sliding motion, which also requires revisit of some basic problems in discrete SMC, involving the existence of sliding mode, reachability and stability of sliding motion

To establish SMC for the discrete-time system, its basic principle is exactly the same as that of continuous-time system, which is also divided into two basic issues:

First: select a switching function $S(k)$ to ensure that the sliding mode is globally asymptotically stable;

Second: calculate the sliding mode controller $u(k) = u^{\pm}(k)$, so that all movements can be reached onto the switching surface $S(k) = 0$ in a finite time.

As mentioned above, the arrival conditions in the discrete-time case are mostly extended from the continuous-time case, and the following arrival conditions have been proposed.

The inequality type:

$$[s(k+1) - s(k)]s(k) \leq 0,$$

$$| s(k+1) | \leq | s(k) |,$$

$$V(k+1) - V(k) \leq 0, \ V(k) = \frac{1}{2}s^2(k).$$

The discrete-time exponential reaching law type:

$$s(k+1) = (1-qT)s(k) - \epsilon T sgn(s(k)),$$

in which $\epsilon > 0$, $q > 0$ and $qT < 1$.

The discrete-time exponential reaching law condition differs from the general inequality type of reaching condition in that the influence of sampling period is taken into account. Therefore, there are several advantages for the discrete-time exponential reaching law:

- The system has good quality during the reaching phase, and the parameters ϵ and q in the reaching law can be adjusted;
- The size of the switching band can be calculated;
- Solving the SMC problem becomes more simple;
- The reaching condition in the form of equation gives equation type of the variable structure control, which is easy to impelment in the design process.

Even SMC has the advantages such as fast response, insensitivity to parameter variations and complete rejection of matched external disturbances, and shows strong robustness to external noise interference and parameter perturbation. However, SMC also shows its main disadvantage (chattering effect) in practical application due to the hysteresis of the switching device; the actual sliding mode cannot always remain on the switching manifold, which may be for the following reasons:

1. Space-lag switch
   Switch lag corresponds to the existence of dead zone of state variables in the state space. Therefore, the consequence is that constant amplitude wave will be superimposed on the sliding surface.
2. Time-lag switch
   In the vicinity of sliding surface, due to the time lag of the switch, the exact change of the control effect on the state is delayed for a period of time, and since the amplitude of the control quantity decreases gradually, it shows a decaying triangular wave on the sliding surface.
3. The effect of physical inertia
   Since the energy and acceleration of the system are finite, and the inertia of the system always exits, the control switch is always accompanied by a time lag, which is consistent with the time lag in effect.
4. Chattering generated by the characteristics of the discrete system.
   The sliding mode of discrete-time system itself is a "quasi-sliding mode"; its switching action does not occur on the sliding surface, but on the surface of a conical body whose vertex is the origin point.

In addition, it is well known that the chattering cannot be eliminated, so we need to select appropriate gain parameters to reduce the impact of chattering when designing the sliding mode controller. At present, there are a lot of researches on chattering phenomenon in SMC worldwide, and some conclusions are presented as follows: (1) *quasi-sliding mode method* mainly includes the continuous function approximation method and boundary-layer design method; in order to smooth the switching signal sgn($s(t)$), a continuous sigmoid function was often proposed [5,6]. In Ref. [7], nonadaptive chattering-free sliding mode controllers were proposed by assuming that the norm of the derivative of the sliding surface was upper-bounded; (2) *reaching law methods,* such as using the linear approach law and exponential approach law [8]; (3) *the filtering method,* which enables the control signal of the system passing through the filter for smooth filtering; for instance, in Ref. [9], a gain-scheduled SMC scheme was proposed for tracking control tasks of multilink robotic manipulators, in which two classes of low-pass filters were introduced to work concurrently for the purpose of acquiring equivalent control in order to reduce the switching gains; (4) *the interference observer method,* using the interference observer to estimate external perturbations and uncertainties, then to compensate them [10,11]; (5) *the dynamic sliding mode method* – in order to obtain the essentially time-continuous dynamic SMC law, the switching function used in the conventional SMC was reconstructed through a differential process [12,13]; (6) *fuzzy methods* – one is based on experience, using fuzzy logic to realize self-adjustment of SMC parameters, and the other way is based on the universal approximation property of fuzzy models [14,15]. Other methods to reduce the chattering effect include the neural network, genetic algorithm optimization and switching gain reduction [16–19].

1.1.2 Implementation of SMC

Generally, a conventional SMC design involves two steps: **Step 1.** Designing a sliding surface $S(x) = 0$ such that the dynamics restricted to the sliding surface has the desired properties, such as stability, disturbance rejection capability and tracking; **Step 2.** Designing a discontinuous feedback controller $u(t)$ such that the system-state trajectories can be attracted onto the designed sliding surface in a finite time and maintained on the sliding surface for all subsequent time periods. Figure 1.1 plots the implementation of an ideal SMC process, which includes two phases – one is the ***reaching phase:*** from initial state x_0 to point A on hyperplane – and the other is the ***sliding motion phase:*** from point A to the equilibrium point O.

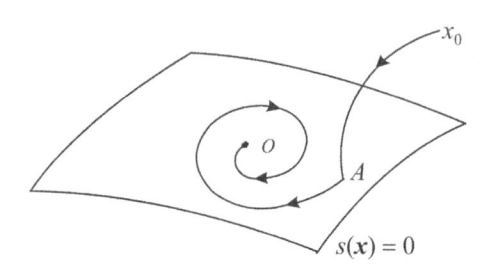

FIGURE 1.1   Implementation of an ideal SMC.

1.1.2.1 Sliding Surface Design

The design of sliding surface is an important part of SMC, which determines the dynamic quality of the obtained sliding mode to a large extent. For different SMC strategies, different switching surfaces can be proposed, such as linear sliding surfaces [20], time-varying sliding surfaces [21], and integral sliding surfaces [22]. In terminal SMC [23], a nonlinear function was introduced in the sliding surface design to realize that the tracking error of the system in sliding mode converged to zero in finite time. In global robust SMC [24], a class of dynamic switching functional is designed to ensure that the system has dynamic sliding quality during the whole process; that is, the reaching motion is eliminated and the system shows strong robustness in the whole response process. The integral sliding surface method can reduce steady-state error of the system and realize global robustness of the system. In view of the aforementioned advantages, these sliding surface design schemes have been widely used in control analysis of various systems.

1.1.2.2 Sliding Mode Controller Design

Another important part of SMC is the design of control law, that is, to find the appropriate controller such that the system meets at least the three properties (the three points in continue-time SMC). So far, with the extensive development of SMC theory, it has been widely applied in the field of robotics and aerospace. The design of sliding mode controller can also be roughly divided into the following categories: (1) the SMC of discrete-time system [25,26] – the main problems to be considered are the chattering effect and large gain feedback; (2) the SMC of mismatched uncertain system [27,28]; (3) the SMC of time-delay systems [29,30]; (4) the SMC of nonlinear system [31,32]; (5) the SMC of stochastic systems [33,34], and other types of complex systems. In terms of control methods, it is worth noting the following hot research fields: (1) the synthesis of adaptive SMC strategy [35,36]; (2) the construction of the sliding mode observer [37,38]; (3) design of fuzzy sliding mode controller [39,40]; (4) design of neural sliding mode controller [41,42]; and (5) the construction of high-order SMC [43,44]. Similarly, with the development of control technology and computer technology, the back-stepping method [45], the linear matrix inequality (LMI) technology [46] and the S-function method [47] have also rapidly promoted the development of SMC.

1.2 SEMI-MARKOVIAN JUMP SYSTEMS

1.2.1 Review of Markovian Jump Systems

With the fast development of modern industrial engineering, the automatic control theory has been applied, which roughly has gone through the following three stages: classical control theory, modern control theory and large-system control theory. On the contrary, the fast development of automatic control theory also promoted the development of modern industrial engineering. In modern industrial engineering, the high demands for modeling complex system with more accuracy and precision are increasing. For example, some abrupt changes in the system, such as dynamic abrupt changes, parameter shifting or environmental noise changes, need to be described by more appropriate mathematical models. In general, these internal or external uncertainties

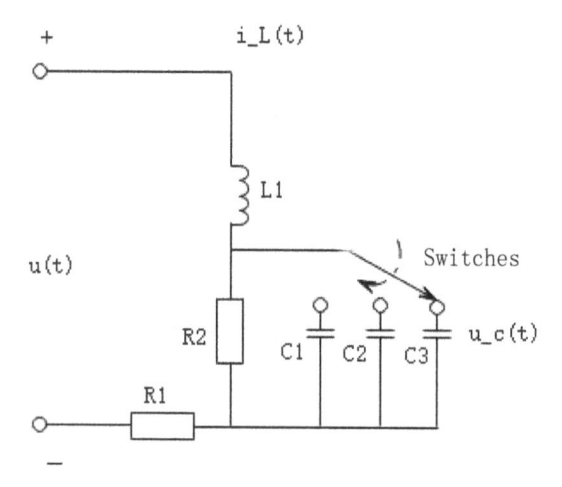

FIGURE 1.2 RLC series circuit.

cannot be described by traditional state-space models, so experts and scholars introduced stochastic processes and stochastic variables. Among the models, the application of stochastic jumping system to model such a physical model with multimode property or an intelligent control system with multicontroller switches has attracted great attention. Particularly, such kinds of stochastic system theory and stochastic control methods have been widely used in all walks of life, such as the network communications, power systems and chemical engineering and many other engineering fields.

Consider an electrical circuit illustrated in Figure 1.2, in which the position of the switch $r(t)$ changes randomly among the capacitors with three states $S = \{1,2,3\}$.

Accordingly, let $x_1(t) = u_c(t)$ and $x_2(t) = i_L(t)$ as state vectors, where $u_c(t)$ and $i_L(t)$ are the voltage of the capacitor and the current of the inductor, respectively. So the model in Figure 1.2 can be represented by the following dynamics:

$$L\frac{di_L(t)}{dt} + u_c(t) + R_1 i_L(t) = u(t),$$

$$C(r_t)\frac{du_c(t)}{dt} + \frac{u_c(t)}{R_2} = i_L(t).$$

By defining $x(t) = [x_1^T(t)\ \ x_2^T(t)]^T$, then it obtains the state-space system in the form:

$$\dot{x}(t) = \begin{bmatrix} -\dfrac{1}{C(r_t)R_2} & \dfrac{1}{C(r_t)} \\[3mm] -\dfrac{1}{L} & -\dfrac{R_1}{L} \end{bmatrix} x(t) + \begin{bmatrix} 0 \\[2mm] \dfrac{1}{L} \end{bmatrix} u(t).$$

The above state-space equation gives the form of a physical system with multimode property modeled by a stochastic system.

As an important class of stochastic hybrid systems, Markovian jumping systems (MJSs) have attracted the attention of many scholars due to their extensive modeling applicability. Moreover, a large number of research results have been extended to various social production processes, such as power system, manufacturing system, communication system, control of nuclear power plant, control of aircraft and wireless servo control, etc. [48–51]. Since Krasovskii and Lidskii [52] proposed the MJSs to describe the stochastic jumping system, a large number of scholars have given many important theoretical achievements on the analysis of various performance indexes and comprehensive control of such systems. These theoretical achievements not only promoted the development of control theory, but also made great contributions to the improvement of social and economic benefits.

Definition 1.1 [53]

A random sequence $\{X(t, n \geq 0), t \geq 0\}$ is a Markov chain if for all, $i_0, i_1 \cdots, i_n, i_{n+1} \in S, n \in N$ and $\Pr\{X_{n+1} = i_{n+1} \mid X_0 = i_0, X_1 = i_1, \cdots, X_n = i_n\} > 0$, it satisfies

$$\Pr\{X_{n+1} = i_{n+1} \mid X_0 = i_0, X_1 = i_1, \cdots, X_n = i_n\} = \Pr\{X_{n+1} = i_{n+1} \mid X_n = i_n\}. \quad (1.2)$$

Definition 1.2 [54]

A Markov chain $\{X(t, n \geq 0), t \geq 0\}$ is homogeneous if the probabilities (1.2) do not depend on n and is non-homogeneous in the other cases.

On the complete probability space $(\Omega, \mathcal{F}, \mathcal{P})$, consider the following linear MJS:

$$\begin{cases} \dot{x}(t) = A(r_t)x(t) + B(r_t)u(t) \\ x(t) = \phi(t), \end{cases} \quad (1.3)$$

where $x(t)$ is the system-state vector, $A(r_t)$ and $B(r_t)$ are the system matrices with compatible dimensions that depend on r_t. $\{r_t, t \geq 0\}$ is a continuous-time homogeneous Markov process taking values in a finite space $S = \{1, 2, \cdots, s\}$. The evolution of the Markov process $\{r_t, t \geq 0\}$ is governed by the following probability transitions:

$$\Pr\{r_{t+h} = j \mid r_t = i\} = \begin{cases} \pi_{ij}h + o(h), & i \neq j, \\ 1 + \pi_{ii}h + o(h), & i = j, \end{cases}$$

where $h > 0$ and $\lim_{h \to 0} o(h) / h = 0$, $\pi_{ij} > 0, i \neq j$, is the transition rate from mode i at time t to mode j at time $t + h$, and $\pi_{ii} = -\sum_{j \neq i} \pi_{ij} < 0$ for all $i \in S$.

Definition 1.3 [55,56]

For the linear SMJ (1.3), for all initial conditions x_0, r_0,
  I: if it satisfies that

$$\lim_{t \to +\infty} \mathbf{E}\left\{ \int_0^t \| x(s) \|^2 \, ds \mid x_0, r_0 \right\} < +\infty,$$

then the system (1.3) is said to be stochastically stable;
  II: if it satisfies that

$$\lim_{t \to +\infty} \mathbf{E}\left\{ \| x(t, x_0, r_0) \|^2 \right\} = 0,$$

then the system (1.3) is said to be mean-square asymptotically stable;
  III: if for arbitrary parameters $a > 0$ and $b > 0$, it satisfies that

$$\mathbf{E}\left\{ \| x(t, x_0, r_0) \|^2 \right\} < a \| x_0 \|^2 e^{-bt},$$

then the system (1.3) is said to be mean-square exponentially stable.
  IV: if it satisfies that

$$\Pr\left\{ \lim_{t \to +\infty} \mathbf{E}\left\{ \| x(t, x_0, r_0) \| \right\} = 0 \right\} = 1,$$

then the system (1.3) is said to be almost surely (asymptotically) stable.

Lemma 1.1 [57]

The linear MJS (1.3) is stochastically stable if there exists positive-definite matrix P_i such that the following condition is satisfied

$$P_i A_i + A_i^T P_i + \sum_{j=1}^{s} \pi_{ij} P_j < 0.$$

1.2.2 DESCRIPTION OF SEMI-MARKOVIAN JUMP SYSTEMS

It is worth noting that modeling stochastic jumping systems using Markov stochastic processes or Markov chains has great advantages. However, in the current pace of society and the speed of technological innovation, especially the rapid development of computer technology combined with artificial intelligence, the needs on the suitability of modeling a physical system are more rigorous. In traditional Markov models, the transition rate induced by the Markov stochastic process is a constant, which originates from the fact that the dwell time of its modes subjects to the memoryless exponential distribution. However, the unique exponential distribution of dwell time also greatly limits the effectiveness of MJSs in specific fields. As a result, some of the theoretical and algorithmic research results for MJSs cannot be applied to certain social industrial production effectively. Recently, the stochastic jumping system using semi-Markovian parameter has attracted the attention of many scholars [58,59], because the semi-Markovian jump systems (S-MJSs) relax the restriction that the distribution of mode dwell time obeys exponential distribution but obeys more general distribution, such as the Gaussian distribution [60] and Weibull distribution [61], which further expands the analysis and synthesis of physical system. From the point view of modeling stochastic systems, the MJSs and the S-MJSs have great similarity that the theoretical research of the S-MJSs can borrow the results and technical methods applied in MJSs for a certain extent. However, the transition rate (TR) of an S-MJS is time-dependent, that is, time-varying. Therefore, the main attention is focused on the influence of the time-varying TRs that play on the whole system performance indexes and on the mechanism of control design. At present, compared with the MJSs, there does not exist a set of mature and consistent theoretical basis for the study of the S-MJSs. This book tries to make some efforts for it.

Generally, in continuous-time stochastic jumping systems, the sojourn time is the time duration between the two jumps. The sojourn time h is a random variable following continuous probability distribution F, for instance; F is an exponential distribution in the MJSs. Correspondingly, the transition rate $\lambda_{ij}(h)$, which is largely determined by F, refers to the speed that the system jumps from mode i to mode j. For example, if F is an exponential distribution, then $\lambda_{ij}(h) \equiv \lambda_{ij}$ is a constant, which derives from the memoryless property of the exponential distribution. As pointed in Ref. [62], only the exponential distribution among the continuous-time probability distributions pertains the memoryless property, which indicates that the jump speed of the stochastic process is independent of the past. Therefore, using MJSs to model stochastic system requires that the transition rate of the system is independent of the past. Due to the exponential distribution for the sojourn time, the transition rate from mode i to mode j is constant, that is,

$$\lambda_{ij} = \frac{f_{ij}(t)}{1 - F_{ij}(t)} = \frac{\lambda_{ij} e^{-\lambda_{ij} t}}{1 - (1 - e^{-\lambda_{ij} t})} = \lambda_{ij},$$

in which $f_{ij}(t)$ is a probability distribution function from mode i to mode j, and $F_{ij}(t)$ is the corresponding cumulative distribution function.

Regarding the semi-Markovian process, for example, the probability distribution of sojourn time follows Weibull distribution: for a given shape parameter $\alpha > 0$ and a scale parameter $\beta > 0$, the probability distribution function is in the form

$$f(h) = \begin{cases} \dfrac{\alpha}{\beta^\alpha} h^{\alpha-1} \exp[-(\dfrac{h}{\beta})^\alpha], & h \geq 0, \\ \\ 0, & h < 0, \end{cases}$$

The cumulative distribution function is known as

$$F(h) = \begin{cases} 1 - \exp[-(\dfrac{h}{\beta})^\alpha], & h \geq 0, \\ \\ 0, & h < 0, \end{cases}$$

Therefore, the transition rate function $\lambda(h)$ can be computed as

$$\lambda(h) = \frac{f(h)}{1 - F(h)} = \frac{\alpha}{\beta^\alpha} h^{\alpha-1}.$$

As a result of the relaxation, the memoryless property in the MJSs does not pertain in the S-MJSs, so the transition rate is not a constant in the S-MJS. This poses the main technical difficulty for the stochastic stability analysis for the S-MJSs. Also, we can see that MJSs are a special kind of S-MJSs with $\alpha = 1$.

On the complete probability space $(\Omega, \mathcal{F}, \mathcal{P})$, consider the following linear S-MJS:

$$\begin{cases} \dot{x}(t) = A(r_t)x(t) + B(r_t)u(t) \\ x(t) = \phi(t), \end{cases} \qquad (1.4)$$

where $x(t)$ is the system-state vector, and $A(r_t)$ are the system matrices with compatible dimensions that depend on r_t. $\{r_t, t \geq 0\}$ is a continuous-time homogeneous semi-Markovian process taking values in a finite space $\mathcal{S} = \{1, 2, \cdots, s\}$.

Definition 1.4 [63]

The evolution of the semi-Markovian process $\{r_t, t \geq 0\}$ is governed by the following probability transitions:

$$\Pr\{r_{t+h} = j \mid r_t = i\} = \begin{cases} \pi_{ij}(h)h + o(h), & i \neq j, \\ 1 + \pi_{ii}(h)h + o(h), & i = j, \end{cases}$$

where $h > 0$ and $\lim_{h \to 0} o(h)/h = 0$, $\pi_{ij}(h) > 0, m \neq n$, is the transition rate from mode m at time t to mode n at time $t + h$, and $\pi_{ii}(h) = -\sum_{j \neq i} \pi_{ij}(h) < 0$ for all $i \in \mathcal{S}$.

In addition, for the system (1.4), we have the following lemma.

Lemma 1.2 [63]

The linear S-MJS (1.4) is stochastically stable if there exists positive-definite matrix P_i such that the following condition is satisfied

$$P_i A_i + A_i^T P_i + \sum_{j=1}^{s} \pi_{ij}(h)P_j < 0.$$

Based on the above lemma, a lot of research results have been proposed. The literature [64,65] proposed the stability analysis for a class of phase-type S-MJS, in which the sojourn time obeyed the phase-type distribution. The stochastic stability and robust-state feedback stabilization of S-MJSs were discussed by LMI method in Ref. [66].

In Ref. [63], the time-varying transition rates of the system were divided into M equal parts by using the segmentation method, and a new criterion for the stochastic stability of the system was given, which greatly reduced the conservatism of the conclusion. In Refs. [67,68], the authors studied the generalized moment stability and stochastic stabilization of linear S-MJSs, respectively. In Ref. [69], Li and Shi systematically proposed the stochastic stability analysis, the estimated state-based SMC design, output feedback quantitative control analysis, filtering design and fault diagnosis of the S-MJSs. Zhang [70] studied the stochastic stability and stabilization of linear discrete-time S-MJSs by using the semi-Markovian kernel method. Through the matrix spectral radius analysis, the stochastic stability analysis of the positive S-MJSs was given in Ref. [71]. In Ref. [72], the SMC approach was applied to study the passive control of S-MJSs in the presence of actuator failure, in which the research was generally based on uncertain transition rate. The application of the semi-Markovian model in cluster synchronous traction control of complex dynamic networks was investigated in Ref. [73], etc. Many scholars have made their contributions to the research of SMC for S-MJSs: for example, as discussed in Refs. [74–79], the design of sliding mode observer, adaptive law and adaptive SMC strategy with switching rules were proposed for S-MJSs or S-MJSs with time-varying delay. For other researches of S-MJSs, such as neural network-based filtering design, stochastic synchronization analysis, fault-tolerant control and finite-time control, please refer to [80–87] and references therein.

For S-MJSs, it is difficult to analyze the performance indexes of the system because of time-varying transition rates, which may show high nonlinearity. Some scholars have made in-depth research on this issue. For example, in Ref. [63], the authors set the value of time-varying transition rates upper- and lower-bounded, and then used the known upper and lower bounds to give the stochastic stability criterion, respectively. Following the method, in Ref. [88], the authors linearized the time-varying transition rates by linear combination functions. In Ref. [83], the authors assumed that the dwell time of the system modes obeyed the completely known probability distributions, such as the Gaussian distribution and Weibull distribution, etc., and thus obtained the known transition rate matrix, based on which the criterion conditions for checking stochastic stability were obtained. More generally, in Ref. [89], the authors for the first time gave the stochastic admissibility of generalized S-MJSs with uncertain transition rates and partially unknown transition rates, which laid a solid foundation for subsequent theoretical research and extension of S-MJSs. However, the studies in Ref. [89] still have some deficiencies: for example, how to deal with the stability and stabilization issues when the jumping information from one mode to others is completely unknown. Therefore, the study of the S-MJSs should be further discussed.

1.3 PREVIEW OF THIS BOOK

The contents of this book are organized as follows:

Chapter 1 gives the research background, motivations and research problems of this book, which mainly involve SMC methodologies in continuous-time and discrete-time, application and modeling of MJSs and S-MJSs. A survey is provided on the fundamental theory of the SMC methodologies, which include some basic concepts, sliding surface design, sliding mode controller design and chattering problems. Then, an overview of recent developments of S-MJSs is presented, which includes comparisons with MJSs and latest results on S-MJSs. Finally, we summarize the main contributions of this book and give the outline of this book.

Chapter 2 is devoted to provide further criterion for stochastic stability analysis of semi-Markovian jump linear systems (S-MJLSs), in which more generic transition rates will be studied. As is known, the time-varying TR is one of the key issues to be considered in the analysis of S-MJLS. Therefore, this article is to investigate general cases for the TRs that covered almost all types, especially for the type that the jumping information from one mode to another is fully unknown, which is merely investigated before. By virtue of stochastic functional theory, sufficient conditions are developed to check stochastic stability of the underlying systems via linear matrix inequalities formulation combined with a maximum optimization algorithm. Finally, numerical examples are given to verify the validity and effectiveness of the obtained results.

Chapter 3 addresses the issue of robust fuzzy SMC for continuous-time nonlinear Takagi–Sugeno fuzzy systems with semi-Markovian switching. The focus is on designing a novel fuzzy integral sliding surface without assuming that the input matrices are the same with full column rank and then developing a fuzzy sliding mode controller for stochastic stability purpose. Based on Lyapunov theory, a set of newly developed LMI conditions are established for stochastic stability of the sliding mode dynamics with generally uncertain transition rates, and then extended to where the input matrix is plant-rule-independent, as discussed in most existing literature. Furthermore, finite-time reachability of the sliding surface is also guaranteed by the proposed fuzzy sliding mode control laws. A practical example is provided to demonstrate the effectiveness of the established method numerically.

Chapter 4 is concerned with finite-time SMC of continuous-time S-MJSs with immeasurable premise variables via fuzzy approach. First, an integral sliding surface is constructed based on fuzzy observer. Second, an observer-based SMC law is synthesized to guarantee finite-time reachability of the predefined sliding surface before the prescribed time. Third, through finite-time boundedness analysis, the required boundedness performance is conducted at the reaching phase first and then the sliding motion phase, respectively. Furthermore, sufficient conditions in terms of linear matrix inequalities are established to guarantee the required boundedness performance of the overall closed-loop controlled system during the two phases with generally uncertain transition rates simultaneously. Finally, a practical example is given to show the validity of the established method numerically.

Chapter 5 deals with the issue of observer-based adaptive SMC of nonlinear Takagi–Sugeno fuzzy systems with semi-Markovian switching and immeasurable premise variables. More general nonlinear systems are described in the model since the selections of premise variables are the states of the system. First, a novel integral sliding surface function is proposed on the observer space; then the sliding mode dynamics and error dynamics are obtained in accordance with estimated premise variables. Second, sufficient conditions for stochastic stability with an H_∞ performance disturbance attenuation level γ of the sliding mode dynamics with different input matrices are obtained based on generally uncertain transition rates. Third, an observer-based adaptive controller is synthesized to ensure the finite-time reachability of a predefined sliding surface. Finally, the single-link robot arm model is provided to verify the control scheme numerically.

Chapter 6 proposes a decentralized adaptive SMC scheme for the stabilization of large-scale semi-Markovian jump-interconnected systems, in which dead-zone linearity in the input and unknown interconnections among subsystems are to be dealt with. By designing integral sliding surface for each subsystem, local sliding mode dynamics is obtained, which has good property of dynamics. Based on the LMI technique, sufficient conditions are established for checking the stochastic stability of the sliding mode dynamics under generally uncertain transition rates. The developed local adaptive sliding mode laws not only guarantee finite-time reachability of sliding surface, but also compensate the effects from dead-zone nonlinearity in the input and unknown interconnections among subsystems. Finally, a numerical example is proposed to verify the effectiveness of the control scheme.

Chapter 7 concerns the problem of SMC design for nonlinear switching semi-Markovian jump delayed systems. By choosing a linear switching surface function, we first derive reduced-order sliding mode dynamics. Then, the property of sliding mode dynamics with generally uncertain transition rates is analyzed by checking a set of new conditions. Further, an adaptive SMC law is constructed to ensure the finite-time reaching condition. Finally, practical examples are provided to disseminate the effectiveness of the proposed method numerically.

1.4 SOME USEFUL DEFINITIONS AND LEMMAS

Lemma 1.3 [90]

(Schur complement) Consider a symmetric matrix Q such that

$$Q = \left[\begin{array}{cc} Q_{11} & Q_{12} \\ Q_{21} & Q_{22} \end{array} \right]$$

(i) $Q > 0$ if and only if

$$\begin{cases} Q_{11} > 0, \\ Q_{22} - Q_{12}^T Q_{11}^{-1} Q_{12} > 0 \end{cases}$$

Or

$$\begin{cases} Q_{22} > 0, \\ Q_{11} - Q_{12} Q_{11}^{-1} Q_{12}^T > 0 \end{cases}$$

Lemma 1.4 [91]

Given any real number ε and any square matrix R, the matrix inequality

$$\varepsilon(R + R^T) \le \varepsilon^2 F + R F^{-1} R^T$$

holds for any matrix $F > 0$.

Lemma 1.5 [92]

For any vector $x, y \in \mathbb{R}^n, 0 < P \in \mathbb{R}^{n \times n}, D \in \mathbb{R}^{n \times n_f}, E \in \mathbb{R}^{n \times n_f}$ and $F(t) \in \mathbb{R}^{n_f \times n_f}$ satisfies $F^T(t)F(t) \leq I$; then for any scalar $\varepsilon > 0$ and matrix Q, the following inequalities hold:

1. $2x^T y \leq x^T P x + y^T P^{-1} y$,
2. $Q + DF(t)E + E^T F^T(t)D^T \leq Q + \varepsilon^{-1}DD^T + \varepsilon E^T E$.

Definition 1.5 [93]

Given the Lyapunov functional candidate $V(x(t), r_t, t \geq 0)$, which is twice differentiable on $x(t)$. Then, its infinitesimal operator $\mathcal{L}V(x(t), r_t)$ is defined by

$$\mathcal{L}V(x(t), r_t) = \lim_{\delta \to 0} \frac{\mathbf{E}\{V(x(t+\delta), r_{t+\delta}) \mid x(t), r_t = m)\} - V(x(t), m)}{\delta}.$$

1.5 ABBREVIATIONS AND NOTATIONS

$\mathbb{Z}$ integer

$\mathbb{R}^n$ or $\mathbb{R}^{n \times m}$ field of n-dimensional real vector or $n \times m$ real matrices

Ω sample space

$\mathcal{F}$ σ-algebra

$\mathcal{P}$ probability measurement

$(\Omega, \mathcal{F}, \mathcal{P})$ complete probability space

$\mathbf{E}\{\cdot\}$ expectation operator with respect to probability measures

$\mathcal{L}(\cdot)$ infinitesimal operator of the Lyapunov function

$\mathbf{L}_2[0, +\infty)$ space of square integrable functions on $[0, +\infty)$

$P > 0$ $(P \geq 0)$ P is a symmetric positive (semi-positive) definite matrix

$P - Q$ is a symmetric positive-definite matrix

A^T or A^{-1} transpose of matrix A or inverse of matrix A

I or 0 identity matrix or zero matrix with appropriate dimension

$\varnothing$ empty set

$|\cdot|$ Euclidean vector norm

$\|\cdot\|$ Euclidean matrix norm (spectral norm)

$\|\cdot\|_2$ $\mathbf{L}_2$-norm

$\sum$ sum

$\forall$ for all

$\triangleq$ is defined as

$*$ symmetric terms of a matrix

$\text{Tr}(A)$ trace of the square matrix A

$\text{rank}(A)$ rank of matrix A

$\text{diag}\{a_1, a_2, \ldots, a_n\}$ diagonal matrix with diagonal elements $a_1, a_2, \cdots, a_n$

$\text{sgn}\{\cdot\}$ sign function

$\det(\cdot)$ matrix determinant

$\min(\cdot)$ or $\max(\cdot)$ takes the maximum or minimum value

$\lambda_{\min}(\cdot)$ or $\lambda_{\max}(\cdot)$ minimum or maximum eigenvalue of a matrix

$\text{He}(P)$ defined as $P + P^T$

lim   limit

sup   supremum

inf   infimum

REFERENCES

[1] S. V. Emelyanov, *Variable Structure Control Systems*, Moscow: Nauka (in Russia), 1970.

[2] Y. Itkis, *Control Systems of Variable Structure*. New York: Wiley, 1976.

[3] A. Khalid, J. X. Xu, and X. Yu, On the discrete-time integral sliding-mode control, *IEEE Transactions on Automatic Control*, Vol. 52, No. 4, pp. 709–715, 2007.

[4] S. V. Drakunov, V. I. Utkin, On discrete-time sliding modes, Preprints of the IFAC Symposium on Nonlinear Control System Design, 1989. Pergamon, 1990, pp. 273–278.

[5] J. A. Burton and A. S. I. Zinober, Continuous approximation of variable structure control, *International Journal of Systems Science*, Vol. 17, No. 6, pp. 875–885, 1986.

[6] S. C. Y. Chung and C. L. Lin, A transformed Lure problem for sliding mode control and chattering reduction, *IEEE Transactions on Automatic Control*, Vol. 44, No. 3, pp. 563–568, 1999.

[7] V. Parra-Vega and G. Hirzinger, Chattering-free sliding mode control for a class of non-linear mechanical systems, *International Journal of Robust and Nonlinear Control: IFAC – Affiliated Journal*, Vol. 11, No. 12, pp. 1161–1178, 2001.

[8] W. Gao, Theory and Design Method of Variable Structure Control, Beijing: Science and Technology Press, 1996.

[9] J. Xu, Y. Pan and T. Lee, A gain scheduled sliding mode control scheme using filtering techniques with applications to multilink robotic manipulators, *Journal of Dynamic Systems, Measurement, and Control* , Vol. 122, No. 4, pp. 641–649, 2000.

[10] Y. Kim, Y. Han and W. You, Disturbance observer with binary control theory, *PESC Record. 27th Annual IEEE Power Electronics Specialists Conference. IEEE*, Vol. 2, pp. 1229–1234, 1996.

[11] Y. Eun, J. Kim, K. Kim, et al., Discrete-time variable structure controller with a decoupled disturbance compensator and its application to a CNC servomechanism, *IEEE Transactions on Control Systems Technology*, Vol. 7, No. 4, pp. 414–423, 1999.

[12] A. J. Koshkouei, K. J. Burnham and A. S. I. Zinober, Dynamic sliding mode control design, *IEE Proceedings-Control Theory and Applications*, Vol. 152, No. 4, pp. 392–396, 2005.

[13] M. Zribi, H. Sira-Ramirez and A. Ngai, Static and dynamic sliding mode control schemes for a permanent magnet stepper motor, *International Journal of Control*, Vol. 74, No. 2, pp. 103–117, 2001.

[14] J. Lo and Y. Kuo, Decoupled fuzzy sliding-mode control, *IEEE Transactions on Fuzzy Systems*, Vol. 6, No. 3, pp. 426–435, 1998.

[15] X. Yu, Z. Man and B. Wu, Design of fuzzy sliding-mode control systems, *Fuzzy Sets and Systems*, Vol. 95, No. 3, pp. 295–306, 1998.

[16] K. Jezernik, M. Rodič and B. Curk, Neural network sliding mode robot control, *Robotica*, Vol. 15, No. 1, pp. 23–30, 1997.

[17] Y. Li, K. C. Ng, D. J. Murray-Smith, et al. Genetic algorithm automated approach to the design of sliding mode control systems, *International Journal of Control*, Vol. 63, No. 4, pp. 721–739, 1996.

[18] M. J. Mahmoodabadi, M. Taherkhorsandi, M. Talebipour, et al. Adaptive robust PID control subject to supervisory decoupled sliding mode control based upon genetic algorithm optimization, *Transactions of the Institute of Measurement and Control*, Vol. 37, No. 4, pp. 505–514, 2015.

[19] J. Yang, S. Li and X. Yu, Sliding-mode control for systems with mismatched uncertainties via a disturbance observer, *IEEE Transactions on Industrial Electronics*, Vol. 60, No. 1, pp. 160–169, 2012.

[20] H. H. Choi, An explicit formula of linear sliding surfaces for a class of uncertain dynamic systems with mismatched uncertainties, *Automatica*, Vol. 34, No. 8, pp. 1015–1020, 1998.

[21] F. Harashima, H. Hashimoto and K. Maruyama, Sliding mode control of manipulator with time-varying switching surfaces, *Transactions of the Society of Instrument and Control Engineers*, Vol. 22, No. 3, pp. 335–342, 1986.

[22] Y. Niu, D. W. C. Ho and J. Lam, Robust integral sliding mode control for uncertain stochastic systems with time-varying delay, *Automatica*, Vol. 41, No. 5, pp. 873–880, 2005.

[23] Y. Feng, X. Yu and Z. Man, Non-singular terminal sliding mode control of rigid manipulators, *Automatica*, Vol. 38, No. 12, pp. 2159–2167, 2002.

[24] H. Pang and G. Tang, Global robust optimal sliding mode control for a class of uncertain linear systems, 2008 *Chinese Control and Decision Conference. IEEE*, pp. 3509–3512, 2008.

[25] K. Furuta, Sliding mode control of a discrete system, *Systems & Control Letters*, Vol. 14, No. 2, pp. 145–152, 1990.

[26] X. Su, X. Liu, P. Shi, et al. Sliding mode control of discrete-time switched systems with repeated scalar nonlinearities, *IEEE Transactions on Automatic Control*, Vol. 62, No. 9, pp. 4604–4610, 2016.

[27] D. Ginoya, P. Shendge and S. Phadke, Sliding mode control for mismatched uncertain systems using an extended disturbance observer, *IEEE Transactions on Industrial Electronics*, Vol. 61, No. 4, pp. 1983–1992, 2013.

[28] C. Wen and C. Cheng, Design of sliding surface for mismatched uncertain systems to achieve asymptotical stability, *Journal of the Franklin Institute*, Vol. 345, No. 8, pp. 926–941, 2008.

[29] L. Wu, X. Su and P. Shi, Sliding mode control with bounded L_2 gain performance of Markovian jump singular time-delay systems, *Automatica*, Vol. 48, No. 8, pp. 1929–1933, 2012.

[30] X. Zhao, H. Yang, W. Xia, et al. Adaptive fuzzy hierarchical sliding-mode control for a class of MIMO nonlinear time-delay systems with input saturation, *IEEE Transactions on Fuzzy Systems*, Vol. 25, No. 5, pp. 1062–1077, 2016.

[31] Y. Feng, X. Yu and F. Han. On nonsingular terminal sliding-mode control of nonlinear systems, *Automatica*, Vol. 49, No. 6, pp. 1715–1722, 2013.

[32] J. Yang, J. Wu and A. K. Agrawal, Sliding mode control for nonlinear and hysteretic structures, *Journal of Engineering Mechanics*, Vol. 121, No. 12, pp. 1330–1339, 1995.

[33] L. Wu, Y. Gao, J. Liu, et al. Event-triggered sliding mode control of stochastic systems via output feedback, *Automatica*, Vol. 82, pp. 79–92, 2017.

[34] M. Liu, L. Zhang, P. Shi P, et al., Robust control of stochastic systems against bounded disturbances with application to flight control, *IEEE transactions on Industrial Electronics*, Vol. 61, No. 3, pp. 1504–1515, 2013.

[35] F. Plestan, Y. Shtessel, V. Bregeault, et al., New methodologies for adaptive sliding mode control, *International Journal of Control*, Vol. 83, No. 9, pp. 1907–1919, 2010.

[36] G. Bartolini, A. Ferrara and V. I. Utkin, Adaptive sliding mode control in discrete-time systems, *Automatica*, Vol. 31, No. 5, pp. 769–773, 1995.

[37] H. Li, P. Shi, D. Yao, et al. Observer-based adaptive sliding mode control for nonlinear Markovian jump systems, *Automatica*, Vol. 64, pp. 133–142, 2016.

[38] J. Liu, S. Vazquez, L. Wu, et al. Extended state observer-based sliding-mode control for three-phase power converters, *IEEE Transactions on Industrial Electronics*, Vol. 64, No. 1, pp. 22–31, 2016.

[39] Y. Guo and P. Y. Woo, An adaptive fuzzy sliding mode controller for robotic manipulators, *IEEE Transactions on Systems, Man, and Cybernetics-Part A: Systems and Humans*, Vol. 33, No. 2, pp. 149–159, 2003.

[40] M. Y. Hsiao, T. H. S. Li, J. Z. Lee, et al. Design of interval type-2 fuzzy sliding-mode controller, *Information Sciences*, Vol. 178, No. 6, pp. 1696–1716, 2008.

[41] B. S. Park, S. J. Yoo, J B. Park, et al. Adaptive neural sliding mode control of nonholonomic wheeled mobile robots with model uncertainty, *IEEE Transactions on Control Systems Technology*, Vol. 17, No. 1, pp. 207–214, 2008.

[42] P. Lin, C. F. Hsu, T. T. Lee, et al. Robust fuzzy-neural sliding-mode controller design via network structure adaptation, *IET Control Theory & Applications*, Vol. 2, No. 12, pp. 1054–1065, 2008.

[43] B. Beltran, T. Ahmed-Ali and M. Benbouzid, High-order sliding-mode control of variable-speed wind turbines, *IEEE Transactions on Industrial Electronics*, Vol. 56, No. 9, pp. 3314–3321, 2008.

[44] V. I. Utkin, Discussion aspects of high-order sliding mode control, *IEEE Transactions on Automatic Control*, Vol. 61, No. 3, pp. 829–833, 2015.

[45] Y. Xia, Z. Zhu and M. Fu. Back-stepping sliding mode control for missile systems based on an extended state observer, *IET Control Theory & Applications*, Vol. 5, No. 1, pp. 93–102, 2011.

[46] S. Dadras, S. Dadras and H. R. Momeni, Linear matrix inequality based fractional integral sliding-mode control of uncertain fractional-order nonlinear systems, *Journal of Dynamic Systems, Measurement, and Control*, Vol. 139, No. 11, pp. 1–7 2017.

[47] J. Liu and X. Wang, *Advanced Sliding Mode Control for Mechanical Systems*, Beijing: Springer, 2012.

[48] V. Ugrinovskii and H. R. Pota, Decentralized control of power systems via robust control of uncertain Markov jump parameter systems, *International Journal of Control*, Vol. 78, No. 9, pp. 662–677, 2005.

[49] C. Andrieu, M. Davy and A. Doucet, Efficient particle filtering for jump Markov systems. Application to time-varying autoregressions, *IEEE Transactions on Signal Processing*, 2003, Vol. 51, No. 7, pp. 1762–1770, 2003.

[50] J. Bai and P. Wang, Conditional Markov chain and its application in economic time series analysis, *Journal of Applied Econometrics*, Vol. 26, No. 5, pp. 715–734, 2011.

[51] R. S. Ellis, Large deviations for the empirical measure of a Markov chain with an application to the multivariate empirical measure, *The Annals of Probability*, Vol. 16, No. 4, pp. 1496–1508, 1988.

[52] N. N. Krasovskii, and E. A. Lidskii, Analysis design of controllers in systems with random attributes. *Part I. Automation and Remote Control*, Vol. 22, pp. 1021–1025, 1961.

[53] J. G. Kemeny and J. L. Snell, Markov Chains, New York: Springer-Verlag, 1976.

[54] O. L. D. Valle Costa, M. D. Fragoso and M. G. Todorov, *Continuous-time Markov Jump Linear Systems*, Berlin Heidelberg: Springer Science & Business Media, 2012.

[55] L. Hu, P. Shi and P. M. Frank, Robust sampled-data control for Markovian jump linear system Berlin Heidelberg s, *Automatica*, Vol. 42, No. 11, pp. 2025–2030, 2006.

[56] F. Kozin, On relations between moment properties and almost sure Lyapunov stability for linear stochastic systems, *Journal of Mathematical Analysis and Applications*, Vol. 10, No. 11, pp. 324–353, 1965.

[57] E. K. Boukas, *Stochastic Switching Systems: Analysis and Design*, Boston, MA: Birkhäuser, 2007.

[58] J. Janssen and R. Manca, *Applied Semi-Markov Processes*, Boston, MA: Springer Science & Business Media, 2006.

[59] V. S. Barbu and N. Limnios, *Semi-Markov Chains and Hidden Semi-Markov Models Toward Applications:Their Use in Reliability and DNA Analysis,* New York: Springer Science & Business Media, 2009.

[60] J. Wang, A. Hertzmann and D. J. Fleet, Gaussian process dynamical models, in Y. Weiss *and* B. Schölkopf *and* J. Platt (Eds.), Advances in Neural Information Processing Systems, pp. 1441–1448, 2006.

[61] H. Rinne, *The Weibull Distribution: A Handbook*, Boca Raton: Chapman and Hall/ CRC, 2008.

[62] S. M. Ross, Introduction to Probability Models. London: Academic Press, 2006.

[63] J. Huang and Y. Shi, H∞ state-feedback control for semi-Markov jump linear systems with time-varying delays, *Journal of Dynamic Systems, Measurement, and Control*, Vol. 135, No. 4, ArticleID 041012, 2013.

[64] Z. Hou, J. Luo, P. Shi P, et al. Stochastic stability of Itô differential equations with semi-Markovian jump parameters, *IEEE Transactions on Automatic Control*, Vol. 51, No. 8, pp. 1383–1387, 2006.

[65] Z. Hou, J. Luo and P. Shi, Stochastic stability of linear systems with semi-Markovian jump parameters, *The ANZIAM Journal*, Vol. 46, No. 3, pp. 331–340, 2005.

[66] J. Huang and Y. Shi, Stochastic stability and robust stabilization of semi-Markov jump linear systems, *International Journal of Robust and Nonlinear Control*, Vol. 23, No. 18, pp. 2028–2043, 2013.

[67] S. H. Kim, Stochastic stability and stabilization conditions of semi-Markovian jump systems with mode transition-dependent sojourn-time distributions, *Information Sciences*, Vol. 385, pp. 314–324, 2017.

[68] H. Schioler, M. Simonsen and J. Leth, Stochastic stability of systems with semi-Markovian switching, *Automatica*, Vol. 50, No. 11, pp. 2961–2964, 2014.

[69] F. Li, P. Shi and L. Wu, *Control and Filtering For Semi-Markovian Jump Systems*, Cham: Springer, 2017.

[70] L. Zhang, Y. Leng and P. Colaneri, Stability and stabilization of discrete-time semi-Markov jump linear systems via semi-Markov kernel approach, *IEEE Transactions on Automatic Control*, Vol. 61, No. 2, pp. 503–508, 2016.

[71] M. Ogura and C. F. Martin, Stability analysis of positive semi-Markovian jump linear systems with state resets, *SIAM Journal on Control and Optimization*, Vol. 52, No. 3, pp. 1809–1831, 2014.

[72] B. Jiang, Y. Kao, C. Gao, et al. Passification of uncertain singular semi-Markovian jump systems with actuator failures via sliding mode approach, *IEEE Transactions on Automatic Control*, Vol. 62, No. 8, pp. 4138–4143, 2017.

[73] T. H. Lee, Q. Ma, S. Xu, et al. Pinning control for cluster synchronisation of complex dynamical networks with semi-Markovian jump topology, *International Journal of Control*, Vol. 88, No. 6, pp. 1223–1235, 2015.

[74] F. Li, L. Wu, P. Shi, et al. State estimation and sliding mode control for semi-Markovian jump systems with mismatched uncertainties, *Automatica*, Vol. 51, pp. 385–393, 2015.

[75] Y. Wei, J. H. Park, Qiu J, et al. Sliding mode control for semi-Markovian jump systems via output feedback, *Automatica*, Vol. 81, pp. 133–141, 2017.

[76] Q. Zhou, D. Yao, J. Wang, et al. Robust control of uncertain semi-Markovian jump systems using sliding mode control method, *Applied Mathematics and Computation*, Vol. 286, pp. 72–87, 2016.

[77] W. Qi, J. H. Park, J. Cheng, et al. Robust stabilisation for non-linear time-delay semi-Markovian jump systems via sliding mode control, *IET Control Theory & Applications*, Vol. 11, No. 10, pp. 1504–1513, 2017.

[78] W. Qi, G. Zong and H. R. Karimi. Observer-based adaptive SMC for nonlinear uncertain singular semi-Markov jump systems with applications to DC motor, *IEEE Transactions on Circuits and Systems I: Regular Papers*, Vol. 65, No. 9, pp. 2951–2960, 2018.

[79] B. Jiang, H. R. Karimi, Y. Kao, et al. Reduced-order adaptive sliding mode control for nonlinear switching semi-Markovian jump delayed systems, *Information Sciences*, Vol. 477, pp. 334–348, 2019.

[80] F. Li, P. Shi, L. Wu, et al. Quantized control design for cognitive radio networks modeled as nonlinear semi-Markovian jump systems, *IEEE Transactions on Industrial Electronics*, Vol. 62, No. 4, pp. 2330–2340, 2015.

[81] P. Shi, F. Li, L. Wu, et al. Neural network-based passive filtering for delayed neutral-type semi-Markovian jump systems, *IEEE Transactions on Neural Networks and Learning Systems*, Vol. 28, No. 9, pp. 2101–2114, 2017.

[82] F. Li, L. Wu and P. Shi, Stochastic stability of semi-Markovian jump systems with mode-dependent delays, *International Journal of Robust and Nonlinear Control*, Vol. 24, No. 18, pp. 3317–333, 2014.

[83] Y. Wei, J. H. Park, H.R. Karimi, et al. Improved stability and stabilization results for stochastic synchronization of continuous-time semi-Markovian jump neural networks with time-varying delay, *IEEE Transactions on Neural Networks and Learning Systems*, Vol. 29, No. 6, pp. 2488–2501, 2018.

[84] Y. Wei, J. Qiu, H. R. Karimi, et al. A novel memory filtering design for semi-Markovian jump time-delay systems, *IEEE Transactions on Systems, Man, and Cybernetics: Systems*, Vol. 48, No. 12, pp. 2229–2241, 2018.

[85] L. Chen, X. Huang, and S. Fu, Observer-based sensor fault-tolerant control for semi-Markovian jump systems, *Nonlinear Analysis: Hybrid Systems*, Vol. 22, pp. 161–177, 2016.

[86] K. Liang, M. Dai, H. Shen, et al. $L_2 - L_\infty$ synchronization for singularly perturbed complex networks with semi-Markov jump topology, *Applied Mathematics and Computation*, Vol. 321, pp. 450–462, 2018.

[87] X. Liu, X. Yu, X. Zhou, et al. Finite-time H_∞ control for linear systems with semi-Markovian switching, *Nonlinear Dynamics*, Vol. 85, No. 4, pp. 2297–2308, 2016.

[88] H. Shen, L. Su, and J. H. Park, Reliable mixed H_∞/passive control for T-S fuzzy delayed systems based on a semi-Markov jump model approach. *Fuzzy Sets and Systems*, Vol. 314, pp. 79–98, 2017.

[89] B. Jiang, Y. Kao, H. R. Karimi, et al. Stability and stabilization for singular switching semi-Markovian jump systems with generally uncertain transition rates, *IEEE Transactions on Automatic Control*, Vol. 63, No. 11, pp. 3919–3926, 2018.

[90] F. Zhang, *The Schur Complement and Its Applications*, New York: Springer Science & Business Media, 2006.

[91] J. Xiong, and J. Lam, Robust H_2 control of Markovian jump systems with uncertain switching probabilities, *International Journal of Systems Science*, Vol. 40, No. 3, pp. 255–265, 2009.

[92] Y. Wang, L. Xie, and C. E. D. Souz, Robust control of a class of uncertain nonlinear systems, *Systems & Control Letters*, Vol. 19, No. 2, pp. 139–149, 1992.

[93] S. Meyn and R. Tweedie, *Markov chains and stochastic stability*, Springer Science & Business Media, 2012.

2 Stochastic Stability of Semi-Markovian Jump Systems with Generally Uncertain Transition Rates

2.1 INTRODUCTION

It is known that transition rate is one of the most important roles that regulates the overall dynamic performance of S-MJSs, which distinguishes the dynamics of S-MJSs from that of deterministic switching systems. However, due to the high complexity of real-world circumstance, obtaining the exact knowledge of transition rates seems impossible. Therefore, it is necessary to undertake an analysis of S-MJSs with deficient-mode information. So far, it has witnessed some inspiring works. For example, in Ref. [1], the transition rates were set to follow fully known probability distributions, such as the Gaussian distribution. In Ref. [2], the authors confined the transition rates $\pi_{ij}(h)$ lower- and upper-bounded. And in Ref. [3], a phase-type semi-Markov process was transformed into its associated Markov chain. However, it is noted that all the results proposed in the above literature required information on all transition rates, either full or partial cases. Therefore, the analysis and synthesis of S-MJSs with generally uncertain TRs becomes a hot issue. It has witnessed some results in the field of MJSs: for example, the stability and stabilization problems for a class of continuous-time and discrete-time linear MJS with partly unknown transition probabilities were investigated in Ref. [4], which covers completely known and completely unknown transition probabilities as two special cases. Also in Ref. [5], the stability and stabilization problems for a kind of continuous-time and discrete-time MJS with generally bounded transition rates (probability) were studied. Recently, some new results are proposed in Ref. [6] with a novel method to deal with generally uncertain time-varying transition rates, that is, some transition rates are fully unknown. But in Ref. [6], only two cases for the mode information in the transition rate matrix are investigated, which is far from enough since some important and difficult issues have not been touched yet, especially for the types that transition rates from one mode to others are totally unknown. Although some methods have been proposed in dealing with MJSs with deficiency-mode information [7], the techniques cannot be extended to S-MJSs directly or might be invalid for S-MJSs. So far, to the best of our knowledge, the issue of stochastic stability analysis for S-MJSs with more generic transition rates is an open challenge that has not been addressed, which motivates us to do the research.

Based on the above analysis, this chapter deals with the problem of stochastic stability analysis for linear S-MJSs with generic transition rates, which includes full information, partial information and totally unknown information about the modes. The main contribution is that a set of feasible criteria are established to check the stochastic stability of the considered systems. Particularly, for the case that all transition rates from one mode to others are fully unknown, a feasible estimation method is proposed to deal with this issue. Then, by solving a maximum optimization problem, feasible solutions for the established linear matrix inequality (LMI) conditions can be obtained in the sense of stochastic stability.

2.2 SYSTEM DESCRIPTION

Consider the following continuous-time linear S-MJS fixed on the probability space $(\Omega, \mathcal{F}, \mathcal{P})$:

$$\dot{x}(t) = A(r_t)x(t),$$
$$x(0) = x_0, r(0) = r_0, \tag{2.1}$$

where $x(t) \in \mathbb{R}^n$ is the state vector, $A(r_t)$ is the mode-dependent system matrix with appropriate dimensions, x_0 is the initial condition, and r_0 is the initial mode of the semi-Markov process. $\{r_t, t \geq 0\}$ is a semi-Markov process taking values in a finite set $S = \{1, 2, \ldots, s\}$ and governed by

$$\Pr\{r_{t+h} = j \mid r_t = i\} = \begin{cases} \pi_{ij}(h)h + o(h), & i \neq j, \\ 1 + \pi_{ii}(h)h + o(h), & i = j, \end{cases} \tag{2.2}$$

where $h > 0$ and $\lim_{h \to 0} o(h)/h = 0$, $\pi_{ij}(h) > 0, i \neq j$, is the transition rate from mode i at time t to mode j at time $t+h$, and $\pi_{ii}(h) = -\sum_{j \neq i} \pi_{ij}(h) < 0$ for each $i \in S$. In addition, $A(r_t)$ will be denoted by A_i in the following.

Based on the above discussion, the transition rate matrix for the system (2.1) can be described as

$$\Pi = \begin{bmatrix} \pi_{11}(h) & \pi_{12}(h) & \cdots & \pi_{1s}(h) \\ \pi_{21}(h) & \pi_{22}(h) & \cdots & \pi_{2s}(h) \\ \vdots & \vdots & \ddots & \vdots \\ \pi_{s1}(h) & \pi_{s2}(h) & \cdots & \pi_{ss}(h) \end{bmatrix}. \tag{2.3}$$

Definition 2.1 [2]

The linear S-MJS (2.1) with all modes and all $t \geq 0$ is said to be stochastically stable if there exists a finite positive constant $T(x_0, r_0)$ such that the following inequality holds for any initial condition (x_0, r_0):

$$\mathbf{E} \int_0^t \| \hat{x}(s) \|^2 \, ds \leq \mu^{-1} \mathbf{E} V(\hat{x}(0), r_0).$$

Remark 2.1

Generally, in order to deal with the analysis and synthesis of continuous-time S-MJSs, the following methods have been proposed to tackle the transition rate $\pi_{ij}(h)$:

a. Huang and Shi have proposed some pioneer works [2], in which $\pi_{ij}(h)$ was set lower- and upper-bounded as $\pi_{ij}(h) \in [\underline{\pi}_{ij}, \overline{\pi}_{ij}]$ with $\underline{\pi}_{ij}$ and $\overline{\pi}_{ij}$ being the known lower and upper bounds of the transition rate $\pi_{ij}(h)$, respectively. Further, in order to reduce conservativeness of the proposed stability criterion, the transition rate was further separated into M sections with division points $\pi_{ij,0}, \pi_{ij,1}, \cdots, \pi_{ij,M-1}, \pi_{ij,M}$;

b. Based on the general assumption from Ref. [2] that $\pi_{ij}(h)$ is lower- and upper-bounded, the following conditions are further developed in Ref. [8]:

$$\pi_{ij}(h) = \sum_{k=1}^{K} \zeta_k \pi_{ijk}, \quad \sum_{k=1}^{K} \zeta_k = 1, \zeta_k \geq 0, \tag{2.4}$$

with

$$\pi_{ijk} = \begin{cases} \underline{\pi}_{ij} + (k-1)\dfrac{\overline{\pi}_{ij} - \underline{\pi}_{ij}}{K-1}, & i \neq j, \quad j \in \mathcal{S}, \\[2mm] \overline{\pi}_{ij} - (k-1)\dfrac{\overline{\pi}_{ij} - \underline{\pi}_{ij}}{K-1}, & i = j, \quad j \in \mathcal{S}, \end{cases}$$

c. Particularly, the probability distribution function (PDF) of the sojourn time h staying at one mode was selected to obey certain distribution in Ref. [1], such as the Weibull distribution or Gaussian distribution. Then, it was computed that $\mathbf{E}\{\pi_{ij}(h)\} = \int_0^\infty \pi_{ij}(h) g_i(h) dh$, where $g_i(h)$ is the PDF. Therefore, the mathematical expectation of the transition rate matrix can be defined by

$$\begin{bmatrix} \mathbf{E}\{\pi_{11}(h)\} & \mathbf{E}\{\pi_{12}(h)\} & \cdots & \mathbf{E}\{\pi_{1s}(h)\} \\ \mathbf{E}\{\pi_{21}(h)\} & \mathbf{E}\{\pi_{22}(h)\} & \cdots & \mathbf{E}\{\pi_{2s}(h)\} \\ \vdots & \vdots & \ddots & \vdots \\ \mathbf{E}\{\pi_{s1}(h)\} & \mathbf{E}\{\pi_{s2}(h)\} & \cdots & \mathbf{E}\{\pi_{ss}(h)\} \end{bmatrix} \tag{2.5}$$

d. In view of the above three cases, which all need partial or full information on the TR $\pi_{ij}(h)$, the following TR matrix in Ref. [9] was introduced to better accommodate practical dynamical systems

$$\begin{bmatrix} \pi_{11} + \Delta\pi_{11}(h) & ? & \cdots & \pi_{1s} + \Delta\pi_{1s}(h) \\ ? & ? & \cdots & \pi_{2s} + \Delta\pi_{2s}(h) \\ \vdots & \vdots & \ddots & \vdots \\ ? & \pi_{s2} + \Delta\pi_{s2}(h) & \cdots & \pi_{ss} + \Delta\pi_{ss}(h) \end{bmatrix}. \quad (2.6)$$

This chapter will follow the case (d) and extend the result to more generic cases. Now, revisit the case (d) in Ref. [9]. Practically, for the high complexity of the real-world dynamics, the TRs for S-MJSs are considered to satisfy the following two cases: (I): $\pi_{ij}(h)$ is fully unknown and (II): $\pi_{ij}(h)$ is not exactly known but upper- and lower-bounded. Following [1] that $\pi_{ij}(h) \in [\underline{\pi}_{ij}, \bar{\pi}_{ij}]$, in which $\underline{\pi}_{ij}$ and $\bar{\pi}_{ij}$ are the known real constants representing the lower and upper bounds of $\pi_{ij}(h)$, respectively. Denote $\pi_{ij}(h) \triangleq \pi_{ij} + \Delta\pi_{ij}(h)$, in which $\pi_{ij} = \dfrac{1}{2}(\underline{\pi}_{ij} + \bar{\pi}_{ij})$ and $|\Delta\pi_{ij}(h)| \le \delta_{ij}$ with $\delta_{ij} = \dfrac{1}{2}(\bar{\pi}_{ij} - \underline{\pi}_{ij})$. The TR matrix with three jumping modes may be described as

$$\begin{bmatrix} \pi_{11} + \Delta\pi_{11}(h) & ? & \pi_{13} + \Delta\pi_{13}(h) \\ ? & ? & \pi_{23} + \Delta\pi_{23}(h) \\ ? & \pi_{32} + \Delta\pi_{32}(h) & ? \end{bmatrix}, \quad (2.7)$$

where "?" is the description of unknown TRs. For brevity, $\forall i \in \mathcal{S}$, let $I_i = I_{i,kn} \cup I_{i,ukn}$, where $I_{i,kn}$ and $I_{i,ukn}$ are defined as follows:

$$I_{i,kn} \triangleq \{j : \pi_{ij} \text{ can be determined for } j \in \mathcal{S}\},$$

$$I_{i,ukn} \triangleq \{j : \pi_{ij} \text{ is not known for } j \in \mathcal{S}\}.$$

Basically, the results developed in Ref. [9] have some limitations due to the assumptions that both $I_{i,kn}$ and $I_{i,ukn}$ are not empty. Therefore, only two cases are studied in Ref. [9], that is, $i \in I_{i,kn}$ with $I_{i,ukn} \ne \varnothing$ and $i \in I_{i,ukn}$ with $I_{i,kn} \ne \varnothing$. Here, the most important case to be studied is that what if $i \in I_{i,ukn}$ while $I_{i,kn}$ is empty, that is, $I_{i,kn} = \varnothing$. For instance, the TR matrix with four modes is defined by the following form:

$$\begin{bmatrix} \pi_{11} + \Delta\pi_{11}(h) & \pi_{12} + \Delta\pi_{12}(h) & \pi_{13} + \Delta\pi_{13}(h) & \pi_{14} + \Delta\pi_{14}(h) \\ \pi_{21} + \Delta\pi_{11}(h) & ? & \pi_{23} + \Delta\pi_{23}(h) & ? \\ ? & \pi_{32} + \Delta\pi_{32}(h) & \pi_{33} + \Delta\pi_{33}(h) & ? \\ ? & ? & ? & ? \end{bmatrix}. \quad (2.8)$$

Therefore, based on the above TR matrix, the following cases are needed to be studied:

i. $i \in I_{i,kn}$ and $i \in I_{i,kn}$ is fully known, that is, $I_{i,kn} = S$;

ii. $i \in I_{i,kn}$ and $i \in I_{i,kn}$ is partially known, that is, $I_{i,kn} \neq S$ while $I_{i,kn}$ is also not empty;

iii. $i \in I_{i,ukn}$ and $i \in I_{i,kn}$ is partially known, that is, $I_{i,kn} \neq S$ while $I_{i,kn}$ is also not empty;

iv. $i \in I_{i,ukn}$ and $i \in I_{i,kn}$ is fully unknown, that is, $I_{i,kn} = \varnothing$.

Here, we need to point out that the case (iv) will not exist for all $i \in S$ in the following investigation.

Compared with previous literature, more universal TRs (2.8) will be investigated in this chapter. Especially for the case (iv), which is the most challenging part to be analyzed in the sequel. Therefore, in order to give stochastic stability of semi-Markovian jump linear system (S-MJLS) (2.1) in the sense of case (iv), the following estimation method is proposed:

(iv-i): $i \in I_{i,ukn}$ and $I_{i,kn} = \varnothing$, while there exists $j \in I_{j,kn}$ with $I_{j,kn}$ being not empty for $j \neq i$, such as the first and third rows of TR matrix (2.8). In this case, we define

$$\pi_{ii}(h) = a_i \pi_{jj}(h),$$

where a_i is the estimated parameter to be determined.

(iv-ii): $i \in I_{i,ukn}$ and $I_{i,kn} = \varnothing$, while there is no $j \in I_{j,kn}$ for $j \neq i$, that is, the cases (i) and (ii) do not exist: for example, only the forms of the second and fourth row exist in TR matrix (2.8). In this case, we define

$$\pi_{ij^*}(h) = a_i \pi_{jj^*}(h),$$

where $\pi_{jj^*}(h)$ is the partially known j^*-th TR in the j-th row of TR matrix, and a_i is the estimated parameter to be determined.

To this end, another issue comes that how to determine a_i in order to investigate the stochastic stability of an S-MJLS with TR matrix in form of (2.8). Therefore, to give a comprehensive study of S-MJLS with generic uncertain TRs, at least five cases should be considered, which include the cases (i)–(iii) and the case (iv) that is divided into another two cases (iv-i) and (iv-ii). In the following, an LMI-based formulation and an optimization algorithm are proposed to check the stochastic stability and to obtain the estimated parameters a_i. Before moving on, we denote the following nonempty set $I_{i,kn}$:

$$I_{i,kn} \triangleq \{\ell_{i,1}, \ell_{i,2}, \dots, \ell_{i,o}\} \quad 1 \leq o < s,$$

where $\ell_{i,s}(s \in \{1,2,\dots,o\})$ represents the index of s-th element in the i-th row of the TR matrix.

Therefore, the overall purpose of this chapter is to recommend numerically feasible stochastic stability conditions for the linear S-MJS (2.1) with universal generic TRs.

2.3 STOCHASTIC STABILITY ANALYSIS

Based on the above discussions, the following stochastic stability criterion is proposed for the S-MJSs (2.1) with generic TRs in the presence of the abovementioned five cases.

Theorem 2.1

For the linear S-MJS (2.1), $\pi_{ij}(h)$ depends on sojourn time h, where h is set to 0 when system jumps. Then the system is stochastically stable if there exist $P_i > 0$ and $i \in \mathcal{S}$, such that for all $i \in \mathcal{S}$

$$A_i^T P_i + P_i A_i + \sum_{j=1}^{s} \pi_{ij}(h) P_j < 0. \tag{2.9}$$

Proof: Consider the following Lyapunov functional:

$$V(x(t), r(t)) = x^T(t) P(r_t) x(t). \tag{2.10}$$

Then, according to Definition 1.5 and the method proposed in Ref. [2], we have

$$\mathcal{L}V(x(t), i) = \lim_{r \to 0} \frac{1}{r} \left[\sum_{j=1, j \neq i}^{s} \Pr\{\eta_{t+r} = j \mid \eta_t = i\} x_r^T P_i x_r \right.$$

$$\left. + \Pr\{\eta_{t+r} = j \mid \eta_t = i\} x_r^T P_i x_r - x^T(t) P_i x(t) \right],$$

where $x_\delta \triangleq x(t + \delta)$. For a general distribution of the sojourn time without memory-less property, that is, $\Pr\{\eta_{t+\delta} = j \mid \eta_t = i\} \neq \Pr\{\eta_\delta = j \mid \eta_0 = i\}$, by the conditional probability formula, we have

$$\mathcal{L}V(x(t), i) = \lim_{\delta \to 0} \frac{1}{\delta} \left[\sum_{j=1, j \neq i}^{s} \frac{q_{ij}(G_i(h + \delta) - G_i(t))}{1 - G_i(h)} x_\delta^T P_i x_\delta \right.$$

$$\left. + \frac{1 - G_i(h + \delta)}{1 - G_i(h)} x_\delta^T P_i x_\delta - x^T(t) P_i x(t) \right]$$

$$= \lim_{\delta \to 0} \frac{1}{\delta} \left[\sum_{j=1, j \neq i}^{s} \frac{q_{ij}(G_i(h + \delta) - G_i(h))}{1 - G_i(h)} x_\delta^T P_i x_\delta \right.$$

$$+ \frac{1 - G_i(h + \delta)}{1 - G_i(h)} [x_\delta^T - x^T(t)] P_i x_\delta$$

$$+ \frac{1 - G_i(h + \delta)}{1 - G_i(h)} x^T(t) P^T(k) [x_\delta - x(t)]$$

$$\left. - \frac{G_i(h + \delta) - G_i(h)}{1 - G_i(h)} x^T(t) P_i x(t) \right]$$

where $G_i(h)$ is the cumulative distribution function of the sojourn time when the system stays in mode i and q_{ij} is the probability intensity from mode i to mode j. On the other hand, we have that

$$\lim_{\delta \to 0} \frac{(G_i(h+\delta)-G_i(h))}{(1-G_i(h))\delta} = \frac{1}{1-G_i(h)} \lim_{\delta \to 0} \frac{(G_i(h+\delta)-G_i(h))}{\delta} = \pi_i(h),$$

$$\lim_{\delta \to 0} \frac{1-G_i(h+\delta)}{1-G_i(h)} = 1,$$

where $\pi_i(h)$ is the TR of the system jumping from mode i. Therefore,

$$\mathcal{L}V(x(t),i) = \sum_{j=1,j \neq i}^{s} q_{ij}\pi_i(h)x^T(t)P_ix(t) + 2x^T(t)P_i\dot{x}(t) - \pi_i(h)x^T(t)P_ix(t). \quad (2.11)$$

Now, define $\pi_{ij}(h) = \pi_i(h)q_{ij}$ for $j \neq i$ and $\pi_{ii}(h) = -\sum_{j \neq i}\pi_{ij}(h)$. Overall, we have

$$\mathcal{L}V(t) = x^T(t)\Gamma_i x(t). \quad (2.12)$$

in which $\Gamma_i = P_iA_i + A_i^TP_i + \sum_{j=1}^{s}\pi_{ij}(h)P_j$. It is seen from (2.9) that $\Gamma_{i,m} < 0$, Hence, denote a scalar $\mu \triangleq \lambda_{\min}\{\Gamma_i\} > 0$ such that

$$\mathcal{L}V(\hat{x}(t),r_t) \leq -\mu \| \hat{x}(t) \|^2 . \quad (2.13)$$

Therefore, by Dynkin's formula, we get for any $t > 0$

$$\mathbf{E}V(\hat{x}(t),r_t) - \mathbf{E}V(\hat{x}(0),r_0) \leq -\mu\mathbf{E}\int_0^t \| \hat{x}(s) \|^2 \, ds, \quad (2.14)$$

which yields

$$\mathbf{E}\int_0^t \| \hat{x}(s) \|^2 \, ds \leq \mu^{-1}\mathbf{E}V(\hat{x}(0),r_0). \quad (2.15)$$

Then, it is easy to see that the linear S-MJS (2.1) is stochastically stable.
 Now, based on the above results, let's consider the more general cases.

Theorem 2.2

Given the scalar a_i, the linear S-MJS (2.1) is stochastically stable if there exist positive-definite matrices $P_i > 0$, $U_{ij} > 0, V_{ij} > 0, W_{ij} > 0, S_{ij} > 0$ and $T_{ij} > 0$ such that the following conditions hold for each $i \in \mathcal{S}$,

Case I: $i \in I_{i,kn}, I_{i,kn} = S = \{1,2,\dots,s\}$,

$$\begin{bmatrix} \mathcal{A}_i^{11} & \mathcal{A}_i^{12} \\ * & \mathcal{A}_i^{13} \end{bmatrix} < 0, \tag{2.16}$$

Case II: $i \in I_{i,kn}, \forall l \in I_{i,ukn}, I_{i,kn} \triangleq \{\ell_{i,1}, \ell_{i,2}, \dots, \ell_{i,o_1}\}$,

$$\begin{bmatrix} \mathcal{A}_i^{21} & \mathcal{A}_i^{22} \\ * & \mathcal{A}_i^{23} \end{bmatrix} < 0, \tag{2.17}$$

Case III: $i \in I_{i,ukn}, \forall l \in I_{i,ukn}(l \neq i), I_{i,kn} \triangleq \{\ell_{i,1}, \ell_{i,2}, \dots, \ell_{i,o_2}\}$,

$$P_i - P_l \geq 0, \tag{2.18a}$$

$$\begin{bmatrix} \mathcal{A}_i^{31} & \mathcal{A}_i^{32} \\ * & \mathcal{A}_i^{33} \end{bmatrix} < 0, \tag{2.18b}$$

Case IV: $i \in I_{i,ukn}$, there exists $j \neq i$ such that $j \in I_{j,kn}$, for $\forall l \in I_{i,ukn}$,

$$\begin{bmatrix} P_i A_i + A_i^T P_i + a_i \pi_{jj}(P_i - P_l) + a_i \dfrac{\delta_{jj}^2}{4} S_{ii} & a_i(P_i - P_l) \\ * & -a_i S_{ii} \end{bmatrix} < 0, \tag{2.19}$$

Case V: $i \in I_{i,ukn}$, there is no $j \neq i$ such that $j \in I_{j,kn}$, for $\forall l \in I_{i,ukn}(l \neq i)$,

$$P_i - P_l \geq 0, \tag{2.20a}$$

$$\begin{bmatrix} P_i A_i + A_i^T P_i + a_i \pi_{jj*}(P_{j*} - P_l) + \dfrac{\delta_{jj*}^2}{4} T_{ij*} & a_i(P_{j*} - P_l) \\ * & -a_i T_{ij*} \end{bmatrix} < 0, \tag{2.20b}$$

where

$$\mathcal{A}_i^{11} = P_i A_i + A_i^T P_i + \sum_{j=1,j\neq i}^{s} \frac{(\delta_{ij})^2}{4} U_{ij} + \sum_{j=1}^{s} \pi_{ij} P_j,$$

$$\mathcal{A}_i^{12} = [(P_1 - P_i) \ \dots \ (P_{i-1} - P_i) \ (P_{i+1} - P_i) \ \dots \ (P_s - P_i)],$$

$$\mathcal{A}_i^{13} = [-U_{i1} \ \dots \ -U_{i(i-1)} \ -U_{i(i+1)} \ -U_{is}],$$

$$\mathcal{A}_i^{21} = P_i A_i + A_i^T P_i + \sum_{j\in I_{i,kn}} \left[\frac{(\delta_{ij})^2}{4} V_{ij} + \pi_{ij}(P_j - P_l) \right],$$

$$\mathcal{A}_i^{22} = [(P_{\ell_{i,1}} - P_l) \quad \dots \quad (P_{\ell_{i,o1}} - P_l)],$$

$$\mathcal{A}_i^{23} = [-V_{i\ell_{i,1}} \quad \dots \quad -V_{i\ell_{i,o1}}],$$

$$\mathcal{A}_i^{31} = P_i A_i + A_i^T P_i + \sum_{j \in I_{i,kn}} \left[\frac{(\delta_{ij})^2}{4} W_{ij} + \pi_{ij}(P_j - P_l) \right],$$

$$\mathcal{A}_i^{32} = [(P_{\ell_{i,1}} - P_l) \quad \dots \quad (P_{\ell_{i,o2}} - P_l)],$$

$$\mathcal{A}_i^{33} = [-W_{i\ell_{i,1}} \quad \dots \quad -W_{i\ell_{i,o2}}].$$

Proof: Based on the former Theorem 2.1, the linear S-MJS (2.1) is stochastically stable if it holds

$$A_i^T P_i + P_i A_i + \sum_{j=1}^{s} \pi_{ij}(h) P_j < 0.$$

However, the above inequality is not solvable in the sense of nonlinear transition rates. Now, we are to deal with the term $\sum_{j=1}^{s} \pi_{ij}(h) P_j$. Let's consider the following cases:

Case I: $i \in I_{i,kn}$ and $I_{i,kn} = S$.

According to the partition in (2.8), let $\sum_{j=1, j\neq i}^{s} \pi_{ij} = -\pi_{ii}$. Therefore, $\sum_{j=1}^{s} \pi_{ij}(h) P_j$ can be rewritten as

$$\sum_{j=1}^{s} \pi_{ij}(h) P_j = \sum_{j=1}^{s} \pi_{ij} P_j + \sum_{j=1}^{s} \Delta\pi_{ij}(h) P_j$$

$$= \sum_{j=1}^{s} \pi_{ij} P_j + \sum_{j=1, j\neq i}^{s} \Delta\pi_{ij}(h) P_j + \Delta\pi_{ii}(h) P_i$$

$$= \sum_{j=1}^{s} \pi_{ij} P_j + \sum_{j=1, j\neq i}^{s} \Delta\pi_{ij}(h)\left(P_j - P_i\right) \qquad (2.21)$$

$$= \sum_{j=1}^{s} \pi_{ij} P_j + \sum_{j=1, j\neq i}^{s} \left[\frac{1}{2}\Delta\pi_{ij}(h)\left(P_j - P_i\right) + \frac{1}{2}\Delta\pi_{ij}(h)(P_j - P_i) \right]$$

$$\leq \sum_{j=1}^{s} \pi_{ij} P_j + \sum_{j=1, j\neq i}^{s} \left[\frac{\delta_{ij}^2}{4}U_{ij} + (P_j - P_i)U_{ij}^{-1}(P_j - P_i) \right]$$

By applying Schur complement, it is known that (2.16) guarantees the S-MJS (2.1) is stochastically stable in this case.

Case II: $i \in I_{i,kn}$, $I_{i,kn} \neq S$ and $I_{i,kn} \neq \varnothing$.

First, let $\lambda_{i,kn} \triangleq \sum_{j \in I_{i,kn}} \pi_{ij}(h)$. Since $I_{i,ukn} \neq \varnothing$, it satisfies that $\lambda_{i,kn} < 0$. Therefore, $\sum_{l=1}^{s} \pi_{ij}(h)P_i$ can be written as

$$\sum_{j=1}^{s} \pi_{ij}(h)P_j = \left(\sum_{j \in I_{i,kn}} + \sum_{j \in I_{i,ukn}} \right) \pi_{ij}(h)P_j$$

$$= \sum_{j \in I_{i,kn}} \pi_{ij}(h)P_j - \lambda_{i,k} \sum_{j \in I_{i,ukn}} \frac{\pi_{ij}(h)}{-\lambda_{i,k}}P_j. \tag{2.22}$$

It is obvious that $0 \leq \pi_{ij}(h)/-\lambda_{i,k} \leq 1$ $(\forall j \in I_{i,uk})$ and $\sum_{j \in I_{i,ukn}} \frac{\pi_{ij}(h)}{-\lambda_{i,k}} = 1$. Therefore, for $\forall l \in I_{i,uk}$, it holds that

$$P_i A_i + A_i^T P_i + \sum_{j=1}^{s} \pi_{ij}(h)P_j = \sum_{j \in I_{i,ukn}} \frac{\pi_{ij}(h)}{-\lambda_{i,k}} \left[P_i A_i + A_i^T P_i + \sum_{j \in I_{i,kn}} \pi_{ij}(h)(P_j - P_l) \right]. \tag{2.23}$$

For $0 \leq \pi_{ij}(h) \leq -\lambda_{i,k}$, $P_i A_i + A_i^T P_i + \sum_{j=1}^{s} \pi_{ij}(h)P_j < 0$ is equivalent to

$$P_i A_i + A_i^T P_i + \sum_{j \in I_{i,kn}} \pi_{ij}(h)(P_j - P_l) < 0, \quad \text{for} \quad \forall l \in I_{i,ukn}. \tag{2.24}$$

In formula (2.24), it has

$$\sum_{j \in I_{i,kn}} \pi_{ij}(h)(P_j - P_l) = \sum_{j \in I_{i,kn}} \pi_{ij}(P_j - P_l) + \sum_{j \in I_{i,kn}} \Delta\pi_{ij}(h)(P_j - P_l). \tag{2.25}$$

By virtue of Lemma 1.4 and for any $V_{ij} > 0$, it follows that

$$\sum_{j \in I_{i,kn}} \Delta\pi_{ij}(h)(P_j - P_l) = \sum_{j \in I_{i,kn}} \left[\frac{1}{2}\Delta\pi_{ij}(h)(P_j - P_l) + \frac{1}{2}\Delta\pi_{ij}(h)(P_j - P_l) \right]$$

$$\leq \sum_{j \in I_{i,kn}} \left[\frac{(\delta_{ij})^2}{4}V_{ij} + (P_j - P_l)(V_{ij})^{-1}(P_j - P_l)^T \right]. \tag{2.26}$$

Combining (2.22)–(2.26) and applying Schur complement, it can be deduced from (2.17) that the S-MJS (2.1) is stochastically stable in this case.

Case III: $i \in I_{i,ukn}$, and $I_{i,kn} \neq \varnothing$.

Similarly, denote $\lambda_{i,k} \triangleq \sum_{j \in I_{i,kn}} \pi_{ij}(h)$. Since $i \in I_{i,ukn}$, it holds that $\lambda_{i,k} > 0$. Now, $\sum_{j=1}^{s} \pi_{ij}(h)P_j$ can be represented as

$$
\sum_{j=1}^{s} \pi_{ij}(h)P_j = \sum_{j \in I_{i,kn}} \pi_{ij}(h)P_j + \pi_{ii}(h)P_i + \sum_{j \in I_{i,ukn}, j \neq i} \pi_{ij}(h)P_j
$$

$$
= \sum_{j \in I_{i,kn}} \pi_{ij}(h)P_j + \pi_{ii}(h)P_i - (\pi_{ii}(h) + \lambda_{i,k}) \sum_{j \in I_{i,ukn}, j \neq i} \frac{\pi_{ij}(h)P_j}{-\pi_{ii}(h) - \lambda_{i,k}}. \quad (2.27)
$$

It is easily seen that for any $j \in I_{i,ukn}$, it holds $0 \leq \pi_{ij}(h)/-\pi_{ii}(h) - \lambda_{i,k} \leq 1$ and $\sum_{j \in I_{i,ukn}, j \neq i} \frac{\pi_{ij}(h)}{-\pi_{ii}(h) - \lambda_{i,k}} = 1$. Therefore, for $\forall\ l \in I_{i,ukn} (l \neq i)$, it follows

$$
P_i A_i + A_i^T P_i + \sum_{j=1}^{s} \pi_{ij}(h)P_j = \sum_{j \in I_{i,ukn}, j \neq i} \frac{\pi_{ij}(h)}{-\pi_{ii}(h) - \lambda_{i,k}} [P_i A_i + A_i^T P_i + \pi_{ii}(h)(P_i - P_l)
$$

$$
+ \sum_{j \in I_{i,kn}} \pi_{ij}(h)(P_j - P_l)]. \quad (2.28)
$$

Since $0 \leq \pi_{ij}(h) \leq -\pi_{ii}(h) - \lambda_{i,k}$, $P_i A_i + A_i^T P_i + \sum_{j=1}^{s} \pi_{ij}(h)P_j < 0$ is equivalent to

$$
P_i A_i + A_i^T P_i + \pi_{ii}(h)(P_i - P_l) + \sum_{j \in I_{i,kn}} \pi_{ij}(h)(P_j - P_l) < 0. \quad (2.29)
$$

In addition, for $\pi_{ii}(h) < 0$, then (2.29) holds if

$$
\begin{cases} P_i - P_l \geq 0, \\ P_i A_i + A_i^T P_i \end{cases} + \sum_{j \in I_{i,kn}} \pi_{ij}(h)(P_j - P_l) < 0. \quad (2.30)
$$

In view of (2.25) and (2.26), for any $W_{ij} > 0$, it follows

$$
\sum_{j \in I_{i,kn}} \pi_{ij}(h)(P_j - P_l) \leq \sum_{j \in I_{i,kn}} \pi_{ij}(P_j - P_l)
$$

$$
+ \sum_{j \in I_{i,kn}} \left[\frac{(\delta_{ij})^2}{4} W_{ij} + (P_j - P_l)(W_{ij})^{-1}(P_j - P_l)^T \right]. \quad (2.31)
$$

Combining (2.27)–(2.31), it is known that (2.18a) and (2.18b) guarantee the S-MJS (2.1) is stochastically stable in this case.

Case IV: $i \in I_{i,uk}$, and $I_{i,kn} = \varnothing$, while there exists a $j \neq i$ such that $j \in I_{j,kn}$.

In this case, $\pi_{ii}(h)$ is estimated by $a_i\pi_{jj}(h)$. Denote $\lambda_{i,kn} \triangleq \pi_{ii}(h)$. Therefore, $\sum_{l=1}^{s} \pi_{ij}(h)P_i$ can be written as

$$\sum_{j=1}^{s} \pi_{ij}(h)P_j = \pi_{ii}(h)P_i + \sum_{j\in I_{i,ukn}} \pi_{ij}(h)P_j$$

$$= \pi_{ii}(h)P_i - \lambda_{i,k} \sum_{j\in I_{i,ukn}} \frac{\pi_{ij}(h)}{-\lambda_{i,k}}P_j. \tag{2.32}$$

Note that $\sum_{j\in I_{i,ukn}} \pi_{ij}(h) = -\pi_{ii}(h) = -\lambda_{i,k} > 0$. Therefore, for $\forall l \in I_{i,ukn}$, it satisfies that

$$P_i A_i + A_i^T P_i + \sum_{j=1}^{s} \pi_{ij}(h)P_j = \sum_{j\in I_{i,ukn}} \frac{\pi_{ij}(h)}{-\lambda_{i,k}}\left[P_i A_i + A_i^T P_i + \pi_{ii}(h)(P_i - P_l)\right].$$

$$= P_i A_i + A_i^T P_i + \pi_{ii}(h)(P_i - P_l)$$

$$= P_i A_i + A_i^T P_i + a_i\pi_{jj}(h)(P_i - P_l) \tag{2.33}$$

In above formula (2.33), it is true that

$$a_i\pi_{jj}(h)(P_j - P_l) = a_i\pi_{jj}(P_i - P_l) + a_i\Delta\pi_{jj}(h)(P_i - P_l). \tag{2.34}$$

By virtue of Lemma 1.4 and for any $S_{ii} > 0$, it follows that

$$\Delta\pi_{ii}(h)(P_j - P_l) = \left[\frac{1}{2}\Delta\pi_{jj}(h)(P_i - P_l) + \frac{1}{2}\Delta\pi_{jj}(P_i - P_l)\right]$$

$$\leq \left[\frac{(\delta_{jj})^2}{4}S_{ii} + (P_i - P_l)(S_{ii})^{-1}(P_i - P_l)^T\right]. \tag{2.35}$$

Combining (2.32)–(2.35) and applying Schur complement, it can be seen from (2.19) that the S-MJS (2.1) is stochastically stable in this case.

Case V: $i \in I_{i,uk}$, and $I_{i,kn} = \varnothing$, while there is no $j \neq i$ such that $j \in I_{j,kn}$.

In this case, $\pi_{ij^*}(h)$ is estimated by $\pi_{jj^*}(h)$. Similarly, denote $\lambda_{i,k} \triangleq \pi_{ij^*}(h)(j^* \neq i)$. $\sum_{j=1}^{s} \pi_{ij}(h)P_j$ is now rewritten as

$$\sum_{j=1}^{s} \pi_{ij}(h)P_j = \pi_{ij^*}(h)P_{j^*} + \pi_{ii}(h)P_i + \sum_{j\in I_{i,ukn}, j\neq i} \pi_{ij}(h)P_j$$

$$= \pi_{ij^*}(h)P_{j^*} + \pi_{ii}(h)P_i - (\pi_{ii}(h) + \lambda_{i,k}) \sum_{j\in I_{i,ukn}, j\neq i} \frac{\pi_{ij}(h)P_j}{-\pi_{ii}(h) - \lambda_{i,k}}, \tag{2.36}$$

and it is obvious that for any $j \in I_{i,uk}$, it holds $0 \le \pi_{ij}(h)/[-\pi_{ii}(h)-\lambda_{i,k}] \le 1$ and $\sum_{j \in I_{i,uk}, j \ne i} \dfrac{\pi_{ij}(h)}{-\pi_{ii}(h)-\lambda_{i,k}} = 1$. Therefore, for $\forall l \in I_{i,uk}, l \ne i$,

$$P_i A_i + A_i^T P_i + \sum_{j=1}^{s} \pi_{ij}(h) P_j = \sum_{j \in I_{i,ukn}, j \ne i} \frac{\pi_{ij}(h)}{-\pi_{ii}(h)-\lambda_{i,k}} \Big[P_i A_i + A_i^T P_i$$

$$+ \pi_{ii}(h)(P_i - P_l) + \pi_{ij^*}(h)(P_{j^*} - P_l) \Big] \qquad (2.37)$$

For $0 \le \pi_{ij}(h) \le -\pi_{ii}(h)-\lambda_{i,k}, P_i A_i + A_i^T P_i + \sum_{j=1}^{s} \pi_{ij}(h) P_j < 0$ is equivalent to

$$P_i A_i + A_i^T P_i + \pi_{ii}(h)(P_i - P_l) + a_i \pi_{ij^*}(h)(P_{j^*} - P_l) < 0. \qquad (2.38)$$

Since $\pi_{ii}(h) < 0$, (2.38) holds if

$$\begin{cases} P_i - P_l \ge 0, \\ P_i A_i + A_i^T P_i + a_i \pi_{ij^*}(h)(P_{j^*} - P_l) < 0. \end{cases} \qquad (2.39)$$

In (2.39), for any $T_{ij^*} > 0$, it follows

$$\pi_{ij^*}(h)(P_{j^*} - P_l) = \pi_{ij^*}(P_{j^*} - P_l) + \Delta\pi_{ij^*}(h)(P_{j^*} - P_l)$$

$$\le \pi_{ij^*}(P_{j^*} - P_l) + \left[\frac{(\Delta_{ij})^2}{4} T_{ij^*} + (P_{j^*} - P_l)(T_{ij^*})^{-1}(P_{j^*} - P_l)^T \right]. (2.40)$$

Combining (2.36)–(2.40), it can be deduced from (2.20a) and (2.20b) that the S-MJS (2.1) is stochastically stable in this case. In summary, the S-MJS (2.1) is stochastically stable in the sense of Definition 2.1 with generic uncertain TRs from the above analysis. This completes the proof.

Remark 2.2

To better explain the proposed method in **Cases IV** and **V**, let's consider the following two types of transition rate matrices:

$$(\dagger) \begin{bmatrix} \pi_{11} + \Delta\pi_{11}(h) & \pi_{12} + \Delta\pi_{12}(h) & \pi_{13} + \Delta\pi_{13}(h) \\ ? & \pi_{22} + \Delta\pi_{22}(h) & \pi_{23} + \Delta\pi_{23}(h) \\ ? & ? & ? \end{bmatrix};$$

$$(\dagger\dagger) \begin{bmatrix} ? & \pi_{12} + \Delta\pi_{12}(h) & \pi_{13} + \Delta\pi_{13}(h) \\ ? & ? & \pi_{23} + \Delta\pi_{23}(h) \\ ? & ? & ? \end{bmatrix}.$$

In both Case † and Case ††, it is seen that $I_{3,kn}$ is empty. In Case †, the unknown transition rate $\pi_{33}(h)$ can be estimated by the $\pi_{11}(h)$ or $\pi_{22}(h)$, which is the Case IV studied in the above proof. For the Case ††, the unknown transition rate $\pi_{33}(h)$ cannot be estimated by $\pi_{11}(h)$ or $\pi_{22}(h)$ since they are also unknown. However, we can estimate $\pi_{32}(h)$ by $\pi_{12}(h)$, which is the Case V studied above. In Case I, the assumption that

$$-\pi_{ii} = \sum_{j=1,j\neq i}^{s} \pi_{ij}$$ can be realized by sacrificing some precision on $\Delta\pi_{ij}(h)$ since they are

essentially uncertainties. It can also be seen from the above Cases † and †† that these two cases in Theorem 2.2 will not appear simultaneously; hence, when checking the conditions in Theorem 2.2, there are four cases to be checked at most.

Remark 2.3

In checking the stochastic stability conditions in Theorem 2.2, how to determine a_i is an important issue. In order to solve this problem, the following maximum optimization problem is proposed:

Max. $\sum a_i$ subject to feasible $P_i,\ldots,S_{ii},T_{ij^*}$ in conditions (2.18)–(2.20b)

To this end, the stochastic stability of S-MJS (2.1) with generic uncertain TRs can be checked effectively.

2.4 NUMERICAL EXAMPLES

Example 2.1

Consider the linear S-MJS (2.1) with four modes, and the corresponding parameters are given as follows:

$$A_1 = \begin{bmatrix} -1.2 & 0.4 \\ -0.3 & 0 \end{bmatrix}, A_2 = \begin{bmatrix} -1.0 & 0.3 \\ 1.0 & 0.4 \end{bmatrix},$$

$$A_3 = \begin{bmatrix} -1.5 & -0.5 \\ -1.5 & -0.5 \end{bmatrix}, A_4 = \begin{bmatrix} -1.0 & 0.8 \\ -0.3 & -0.3 \end{bmatrix}.$$

Consider the transition rate matrix that regulates the four operation modes given as follows:

$$\begin{bmatrix} -1.5+\Delta\pi_{11}(h) & 0.5+\Delta\pi_{12}(h) & 0.5+\Delta\pi_{13}(h) & 0.5+\Delta\pi_{14}(h) \\ ? & -3.5+\Delta\pi_{22}(h) & 0.5+\Delta\pi_{23}(h) & ? \\ 0.4+\Delta\pi_{31}(h) & 0.9+\Delta\pi_{32}(h) & ? & ? \\ ? & ? & ? & ? \end{bmatrix},$$

in which it denotes that $|\Delta\pi_{ij}(h)| \leq \delta_{ij} = |0.1*\pi_{ij}|$. In the above transition rate matrix, it is clear that $I_{4,kn} = \varnothing$; therefore, $\pi_{44}(h)$ is not known, which can be estimated by $\pi_{11}(h)$ or $\pi_{22}(h)$. Also, it can be seen that the above transition rate matrix covers the studied **Case I, Case II, Case III** and **Case IV**. Therefore, in checking the conditions (2.16)–(2.19), let $\pi_{44}(h) = a_4\pi_{11}(h)$, then solve the maximum optimization problem as discussed in Remark 2.3. By computing, it is found that the allowable maximum value for a_4 subject to feasible solutions of (2.16)–(2.19) is that $a_4 = 0.341$, and other feasible solutions are obtained as follows:

$$P_1 = \begin{bmatrix} 7.4367 & -3.7670 \\ -3.7670 & 18.2142 \end{bmatrix}, P_2 = \begin{bmatrix} 16.5560 & 0.5583 \\ 0.5583 & 24.5176 \end{bmatrix},$$

$$P_3 = \begin{bmatrix} 16.1151 & -8.1720 \\ -8.1720 & 17.6307 \end{bmatrix}, P_4 = \begin{bmatrix} 16.1110 & -8.1726 \\ -8.1726 & 17.6299 \end{bmatrix},$$

$$U_{12} = \begin{bmatrix} 359.0711 & 172.8321 \\ 172.8321 & 251.7369 \end{bmatrix}, U_{13} = \begin{bmatrix} 731.0239 & 1.9023 \\ 1.9023 & 51.3126 \end{bmatrix},$$

$$U_{14} = \begin{bmatrix} 730.9834 & 1.9519 \\ 1.9519 & 51.3360 \end{bmatrix}, V_{22} = \begin{bmatrix} 178.5772 & 62.2602 \\ 62.2602 & 38.3715 \end{bmatrix},$$

$$V_{23} = \begin{bmatrix} 395.5086 & 98.9363 \\ 98.9363 & 26.5289 \end{bmatrix}, W_{31} = \begin{bmatrix} 695.8909 & -24.8672 \\ -24.8672 & 17.5029 \end{bmatrix},$$

$$W_{32} = 10^3 \times \begin{bmatrix} 1.2527 & 0.5773 \\ 0.5773 & 0.2721 \end{bmatrix}, S_{44} = \begin{bmatrix} 281.7277 & -118.3470 \\ -118.3470 & 293.0703 \end{bmatrix}.$$

Example 2.2

Consider the linear S-MJS (2.1) with four modes, and the corresponding parameters are given as follows:

$$A_1 = \begin{bmatrix} -1.0 & -0.8 \\ 0.2 & 1.0 \end{bmatrix}, A_2 = \begin{bmatrix} -1.2 & 0.4 \\ -1.0 & -0.1 \end{bmatrix},$$

$$A_3 = \begin{bmatrix} -2.0 & -0.5 \\ -1.0 & -0.1 \end{bmatrix}, A_4 = \begin{bmatrix} -1.2 & -0.5 \\ 0.5 & -0.3 \end{bmatrix}.$$

Now, consider another TR matrix that relates the four operation modes given as follows:

$$\begin{bmatrix} ? & 0.3+\Delta\pi_{12}(h) & 0.3+\Delta\pi_{13}(h) & 0.4+\Delta\pi_{14}(h) \\ ? & ? & 0.5+\Delta\pi_{23}(h) & 0.1+\Delta\pi_{24}(h) \\ 1.5+\Delta\pi_{31}(h) & 0.8+\Delta\pi_{32}(h) & ? & ? \\ ? & ? & ? & ? \end{bmatrix},$$

in which we also denote $|\Delta\pi_{ij}(h)| \le \delta_{ij} = |0.1*\pi_{ij}|$. In the above TR matrix, it is also clear that $I_{4,kn} = \varnothing$. However, the diagonal elements of other rows are also unknown; therefore, $\pi_{44}(h)$ cannot be estimated by $\pi_{11}(h)$ or $\pi_{22}(h)$. In this case, the off-diagonal elements are available, which are the Case III and Case V studied in Theorem 1. In checking the conditions (2.17), (2.18), (2.20a) and (2.20b), let $\pi_{43}(h) = a_4\pi_{23}(h)$, then solve the maximum optimization problem as discussed in Remark 2.3. By computing, it is found that the allowable maximum value for a_4 subject to feasible solutions of (2.18a), (2.18b), (2.20a) and (2.20b) is that $a_4 = 6.273$, and other feasible solutions are obtained as follows:

$$P_1 = \begin{bmatrix} 568.6822 & 29.7011 \\ 29.7011 & 579.2482 \end{bmatrix}, P_2 = \begin{bmatrix} 699.8520 & 39.7914 \\ 39.7914 & 580.0282 \end{bmatrix},$$

$$P_3 = \begin{bmatrix} 768.2518 & -17.1130 \\ -17.1130 & 638.1229 \end{bmatrix}, P_4 = \begin{bmatrix} 766.4233 & -17.4062 \\ -17.4062 & 638.0746 \end{bmatrix},$$

$$U_{12} = 10^4 \times \begin{bmatrix} 3.3723 & -1.4517 \\ -1.4517 & 1.6283 \end{bmatrix}, U_{13} = 10^4 \times \begin{bmatrix} 3.8143 & -1.8662 \\ -1.8662 & 1.9717 \end{bmatrix},$$

$$U_{14} = 10^4 \times \begin{bmatrix} 2.9185 & -1.2956 \\ -1.2956 & 1.6103 \end{bmatrix}, V_{23} = 10^4 \times \begin{bmatrix} 1.0748 & -0.1272 \\ -0.1272 & 0.2493 \end{bmatrix},$$

$$V_{24} = 10^4 \times \begin{bmatrix} 2.3822 & -1.1489 \\ -1.1489 & 1.0794 \end{bmatrix}, W_{31} = 10^4 \times \begin{bmatrix} 8.4717 & 0.9059 \\ 0.9059 & 1.1878 \end{bmatrix},$$

$$W_{32} = 10^3 \times \begin{bmatrix} 8.6566 & 0.4086 \\ 0.4086 & 1.9351 \end{bmatrix}, T_{43} = \begin{bmatrix} 4.5663 & -2.4484 \\ -2.4484 & 2.2499 \end{bmatrix}.$$

Remark 2.4

Theorem 2.2 extends the stochastic stability theory of S-MJS with time-varying TRs described by polytopic uncertainties to more generic cases, which cannot be checked by traditional methods. Proposed by the above numerical example, all types of TRs have been tested, especially for the **Cases IV** and **V** verified by the given two TR matrices, which demonstrates that the developed stochastic stability criterion is effective and superior to traditional ones. This chapter presents an estimation method to deal with unknown TRs, and the estimated parameters a_i are constant. In the future, more advanced method should be proposed, such as applying adaptive method to deal with time-varying estimated parameters.

2.5   CONCLUSION

This chapter has studied the stochastic stability of S-MJLSs. Different from previous literatures, more generic TRs have been investigated with five cases. By virtue of stochastic Lyapunov functional theory, a set of novel sufficient conditions have been proposed to check the stochastic stability. In the meantime, the estimated parameters of totally unknown TRs can be obtained by solving an optimization problem. In general, the proposed theory for S-MJLSs almost has taken all types of TRs into consideration, which have not been touched before. Finally, numerical examples have been presented to demonstrate the validity of the proposed results.

REFERENCES

[1] Y. Wei, J. H. Park, J. Qiu, et al., Sliding mode control for semi-Markovian jump systems via output feedback, *Automatica*, Vol. 81, pp. 133–141, 2017.

[2] J. Huang and Y. Shi, Stochastic stability of semi-Markov jump linear systems: an LMI approach, *Paper presented at: Proceedings of the 50th IEEE Conference Decision Control European Control Conference*, pp. 4668–4673, 2011.

[3] F. Li, L. Wu, P. Shi and C. C. Lim, State estimation and sliding mode control for semi-Markovian jump systems with mismatched uncertainties, *Automatica*, Vol. 51, pp. 385–393, 2015.

[4] L. Zhang and E. K. Boukas, Stability and stabilization of Markovian jump linear systems with partly unknown transition probabilities, *Automatica*, Vol. 45, No. 2, pp. 463–468, 2009.

[5] X. Li, W. Zhang and D. Lu, Stability and stabilization analysis of Markovian jump systems with generally bounded transition probabilities, *Journal of the Franklin Institute*, https://doi.org/10.1016/j.jfranklin.2020.04.013.

[6] B. Jiang, H. R. Karimi, Y. Kao, et al., Takagi-Sugeno model based event-triggered fuzzy sliding mode control of networked control systems with semi-Markovian switchings, *IEEE Transactions on Fuzzy Systems*, Vol. 28, No. 4, pp. 673–683, 2020.

[7] E. K. Boukas, Stabilization of stochastic nonlinear hybrid systems, *International Journal of Innovative Computing, Information and Control*, Vol. 1, No. 1, pp. 131–141, 2005.

[8] H. Shen, L. Su and J. H. Park, Reliable mixed H∞/passive control for T-S fuzzy delayed systems based on a semi-Markov jump model approach, *Fuzzy Sets and Systems*, Vol. 314, pp. 79–98, 2017.

[9] B. Jiang, Y. Kao, H. R. Karimi and C. Gao, Stability and stabilization for singular switching semi-Markovian jump systems with generally uncertain transition rates, *IEEE Transactions on Automatic Control*, Vol. 63, No. 11, pp. 3919–3926, 2018.

3 Fuzzy Integral Sliding Mode Control of Semi-Markovian Jump Systems

3.1 INTRODUCTION

In the last century, fuzzy logic theory appeared to be an effective way to deal with the synthesis of complex nonlinear systems. Particularly, the fuzzy schemes based on Takagi–Sugeno (T-S) fuzzy models [1] became the most popular and convenient ones due to their conceptual simplicity. In the T-S fuzzy models, the overall model of the system is achieved by fuzzy "blending" of local linear models through membership functions, which then could take full advantages of modern control tools, such as the linear matrix inequality (LMI) techniques. Therefore, the last decade has witnessed a wealth of research efforts on T-S fuzzy systems, and numerous fruitful results have been reported in different aspects, for instance, stability analysis and stabilization, filter design, observability, H_∞ control and adaptive control [2–15], etc.

As mentioned before, fruitful investigations in T-S fuzzy models and control methodologies of complex nonlinear dynamics undoubtfully give rise to fuzzy sliding mode control (SMC) methods, which are developed to better accommodate T-S fuzzy systems. Also, we have witnessed quite a few applications of T-S fuzzy SMC methods in literature and engineering being proposed. The SMC strategy was used to deal with fuzzy systems with matched/mismatched uncertainties or subject to actuator saturation (see for instance [16–22]). In the meanwhile, it is noted the T-S fuzzy SMC schemes employed in the aforementioned papers are based on linear sliding surface. Recently, integral SMC (ISMC) has become very popular due to its advantages that the reaching phase needed in a conventional SMC approach could be eliminated, and sliding mode motion can be implemented from the very beginning of the control action while retaining the order of original models, which further leads to the robustness property that can be ensured throughout the whole system response. As a result, many significant research works regarding ISMC of T-S fuzzy models have been reported. For example, the ISMC design for nonlinear stochastic systems by means of T-S fuzzy approach was investigated in Ref. [23], a dynamic ISMC for stochastic T-S fuzzy systems was proposed in Refs. [24,25], and the adaptive-based or dissipativity-based ISMC methods were implemented in [26–31], etc.. But it is seen that the abovementioned SMC/ISMC methods are mainly focused on normal systems without taking more general stochastic processes, for example, Markovian jumping process, into consideration, not to mention the semi-Markovian jumping process. So the problem of T-S fuzzy ISMC for semi-Markovian jump T-S models has becoming an interesting issue. Unfortunately, the authors have not found much significant results regarding ISMC for T-S fuzzy systems with semi-Markovian switching.

Additionally, we can find some obvious limitations of current research after synthesizing the abovementioned literature. The most typical one is the requirement that different submodels have to share the same input matrix or have the full column rank. Recently, this problem was taken into consideration in Refs. [24] and [28]. In Ref. [24], the authors first proposed a fuzzy dynamic ISMC approach to its T-S fuzzy approximation model, in which an integral sliding surface was proposed with both the system-state vector and control input vector, while the main disadvantage is that the system matched uncertainties and perturbations cannot be eliminated during the sliding motion phase. In Ref. [28], a new ideal of Lie bracket was used to tackle different input matrices in the IMSC design, while the method would be very complicated if semi-Markovian switching process was taken into account in the T-S fuzzy models. Therefore, a critical issue raises that how to tackle the problem of fuzzy ISMC for T-S systems with semi-Markovian switching parameters while overcoming the aforementioned shortages simultaneously? However, to the best of our knowledge, no results have been reported in the literature so far, which motivates to do this study.

Briefly speaking, this chapter concerns with the problem of robust fuzzy ISMC for a class of continuous-time nonlinear semi-Markovian jump T-S fuzzy systems. We will propose a novel integral sliding surface incorporated with transformed input matrices to better accommodate the characteristic of T-S fuzzy models, and establish feasible easy-checking LMI conditions to ensure stochastic stability of the sliding mode dynamics with generally uncertain transition rates (TRs). The main issues to be investigated include the following: (1) the T-S fuzzy ISMC is proposed for semi-Markovian jump T-S fuzzy systems with parameter uncertainties and external disturbances; (2) a novel fuzzy integral sliding surface function is put forward in which nonlinear compensator is adopted to maintain good property of sliding motion; (3) proposing fuzzy SMC laws to ensure finite-time reachability of sliding surface and maintain sliding motion of all submodels despite existence of uncertain transition information and parameter uncertainties.

3.2 SYSTEM DESCRIPTION

Given the probability space $(\Omega, \mathcal{F}, \mathcal{P})$, consider the following nonlinear semi-Markovian jump T-S fuzzy system:

Plant Rule i : IF $\theta_1(t)$ is M_{i1} and $\theta_2(t)$ is M_{i2} and ... and $\theta_p(t)$ is M_{ip}
THEN

$$\begin{cases} \dot{x}(t) = (A_i(r_t) + \Delta A_i(r_t))x(t) + B_i(r_t)(u(t) + f(x(t), r_t)) \\ x(0) = \varphi(0) \end{cases}, \qquad (3.1)$$

where $\theta_1(t), \theta_2(t), \dots, \theta_p(t)$ are the premise variables, and $M_{ij}(i = 1, 2 \dots, r; j = 1, 2 \dots, p)$ are the fuzzy sets. $x(t) \in \mathbb{R}^n$ is the state vector, $u(t) \in \mathbb{R}^m$ is the control input, $A_i(r_t)$ and $B_i(r_t)$ are the system matrices with compatible dimensions. The parameter uncertainty matrix $\Delta A_i(r_t)$ is assumed to be norm-bounded, that is,

$$\Delta A_i(r_t) = E_i(r_t)F_i(r_t)H_i(r_t), \qquad (3.2)$$

where $E_i(r_t)$ and $H_i(r_t)$ are the known real-constant matrices with appropriate dimensions. $F_i(r_t)$ is an unknown matrix with Lebesgue-measurable elements and satisfies $F_i^T(r_t)F_i(r_t) \leq I$.

$f(x(t), r_t)$ are the unknown uncertainties of the plant which is bounded by

$$\| f(x(t), r_t) \| \leq \beta(r_t) \| x(t) \|, \tag{3.3}$$

where $\beta(r_t)$ are the known positive mode-dependent constants.

The stochastic jump process $\{r_t, t \geq 0\}$ is a continuous-time semi-Markovian process taking values in a finite set $S = \{1, 2, \ldots, s\}$ with generator given by

$$\Pr\{r_{t+h} = n \mid r_t = m\} = \begin{cases} \pi_{mn}(h)h + o(h), & m \neq n, \\ 1 + \pi_{mm}(h)h + o(h), & m = n, \end{cases} \tag{3.4}$$

where $h > 0$ and $\lim_{h \to 0} o(h)/h = 0$, $\pi_{mn}(h) > 0, m \neq n$, is the transition rate from mode m at time t to mode n at time $t + h$, and $\pi_{mm}(h) = -\sum_{n \neq m} \pi_{mn}(h) < 0$ for each $n \in S$. The generator matrix is defined by $\Pi = [\pi_{mn}(h)]_{s \times s}$.

By fuzzy blending, that is, adopting a center-average defuzzifier, product-fuzzy inference and a singleton fuzzifier, the overall fuzzy model is inferred as follows:

$$\dot{x}(t) = \sum_{i=1}^{r} h_i(\theta(t))[(A_i(r_t) + \Delta A_i(r_t))x(t) + B_i(r_t)(u(t) + f(x(t), r_t))], \tag{3.5}$$

where $\theta(t) = [\theta_1(t) \; \theta_2(t) \; \cdots \; \theta_p(t)]^T$, $h_i(\theta(t))$ is the membership function given by

$$h_i(\theta(t)) = \frac{\Pi_{j=1}^{p} \mu_{ij}(\theta_j(t))}{\sum_{i=1}^{r} \Pi_{j=1}^{p} \mu_{ij}(\theta_j(t))},$$

in which $\mu_{ij}(\theta_j(t))$ is the grade of membership of $\theta_j(t)$ in $\theta_j(t)$. For all $t > 0$, it is satisfied that $h_i(\theta(t)) \geq 0$ and $\sum_{i=1}^{r} h_i(\theta(t)) = 1$. For simplicity, $r(t) = m$ is used for notation in the following.

Remark 3.1

In this work, $B_{i,m}(i = 1, 2, \ldots, r; m \in S)$ is not assumed to be plant-rule-independent and be of full column rank, which differs from most existing results in the literature. So the system could better accommodate the characteristics of T-S fuzzy models in practice. On the other hand, a more general semi-Markovian switching process is taken into consideration in the T-S fuzzy systems instead of Markovian switching. As is known, different from Markovian jump systems (MJSs), the sojourn time of submodels in semi-Markovian jump systems (S-MJSs) follows a more general distribution, such as Gaussian distribution, Weibull distribution, etc., and the transition rate is generally time-varying instead of being constant. Therefore, the model proposed is much more general in the field.

3.3 MAIN RESULTS

In this section, a distinguished fuzzy integral switching surface will be designed and stochastic stability analysis for the corresponding sliding mode dynamics will be investigated. First, note that very few works were reported on ISMC of T-F fuzzy systems with semi-Markovian switching. In Ref. [32], a time interval sliding surface function was put forward and the input matrices $B_{i,m}$ for different plant rules were assumed to be the same, which are somewhat restrictive and hard to implement in practice. To this end, the following transformation based on Ref. [16] is introduced:

Define

$$\bar{B}_m = \frac{1}{r}\sum_{i=1}^{r} B_{i,m},$$

satisfying the rank constraint that rank $(\bar{B}_m) = m$, that is, $\bar{B}_m$ has the full column rank. Further, let

$$V_m = \frac{1}{2}[\bar{B}_m - B_{1,m} \quad \bar{B}_m - B_{2,m} \quad \cdots \quad \bar{B}_m - B_{r,m}],$$

$$U(h) = diag\{1 - 2h_1(\theta(t)), 1 - 2h_2(\theta(t)), \ldots, 1 - 2h_r(\theta(t))\},$$

$$W = [I \quad I \quad \cdots \quad I]^T.$$

Therefore,

$$\bar{B}_m + V_m U(h)W = \bar{B}_m + \frac{1}{2}[(\bar{B}_m - B_{1,m})(1 - 2h_1(\theta(t))) + \cdots$$

$$+ (\bar{B}_m - B_{r,m})(1 - 2h_r(\theta(t)))]$$

$$= \bar{B}_m + \frac{1}{2}\bar{B}_m[(1 - 2h_1(\theta(t))) + \cdots + (1 - 2h_r(\theta(t)))]$$

$$- \frac{1}{2}[B_{1,m}(1 - 2h_1(\theta(t))) + \cdots + B_{r,m}(1 - 2h_r(\theta(t)))] \qquad (3.6)$$

$$= \bar{B}_m + \frac{1}{2}\bar{B}_m(r - 2\sum_{i=1}^{r} h_i(\theta(t))) - \frac{1}{2}\sum_{i=1}^{r} B_{i,m} + \sum_{i=1}^{r} h_i(\theta(t))B_{i,m}$$

$$= \sum_{i=1}^{r} h_i(\theta(t))B_{i,m}.$$

Thus, the T-S fuzzy system (3.5) is further rewritten as:

$$\dot{x}(t) = \sum_{i=1}^{r} h_i\big(\theta(t)\big)\Big[\big(A_{i,m} + \Delta A_{i,m}\big)x(t) + \big(\bar{B}_m + \Delta\bar{B}_m\big)(u(t) + f_m(x,t))\Big], \quad (3.7)$$

where $\Delta\bar{B}_m = V_m U(h)W$, in which it holds $U^T(h)U(h) \le I_{r\times n}$.

Remark 3.2

If $B_{i,m} = B_{j,m}(i,j = 1,2,\ldots,r)$, then $V_m = 0$, that is, $\Delta\bar{B}_m = 0$, which is the case that has been studied in most existing literature. So, the investigation here is more general since it does not assume that all $B_{i,m}$ have the full column rank. The demerit of this decomposition could be the possibility that $\sum_{i}^{r} B_{i,m} = 0$, which is of course excluded in this investigation because we want to provide a potential way to study T-S fuzzy models with plant-rule-dependent input matrices.

3.3.1 SLIDING SURFACE DESIGN

In this part, to better accommodate the characteristic of T-S fuzzy systems, the following fuzzy integral switching surface function is proposed:

$$s(t) = Gx(t) - \int_0^t \sum_{i=1}^{r} h_i(\theta(s))G(A_{i,m} + \bar{B}_m K_{i,m})x(s)ds$$

$$- \int_0^t G\Delta\bar{B}_m(u(s) + v(s))ds, \quad (3.8)$$

where $G \in \mathbb{R}^{m\times n}$ is chosen such that $G\bar{B}_m$ is nonsingular, $K_{i,m} \in \mathbb{R}^{m\times n}$ is adopted to stabilize the sliding mode dynamics, so it is selected such that $A_{i,m} + \bar{B}_m K_{i,m}$ is Hurwitz, and $v(t)$ is designed to attenuate the effect of unknown uncertainties $\Delta\bar{B}_m f_m(x,t)$.

Based on the systems (3.7) and (3.8), we have

$$\dot{s}(t) = \sum_{i=1}^{r} h_i(\theta(t))G\Big[\Delta A_{i,m}x(t) + \bar{B}_m(u(t) + f_m(x,t))$$

$$+ \Delta\bar{B}_m(f_m(x,t) - v(t)) - \bar{B}_m K_{i,m}x(t)\Big]. \quad (3.9)$$

When the state trajectories of the system (3.7) reached onto the sliding surface, that is, $s(t) = 0$, $\dot{s}(t) = 0$. By $\dot{s}(t) = 0$, we have the equivalent control variable as

$$u_{eq}(t) = \sum_{i=1}^{r} h_i(\theta(t))[K_{i,m} - (G\bar{B}_m)^{-1}G\Delta A_{i,m}]x(t)$$

$$- (G\bar{B}_m)^{-1}G\Delta\bar{B}_m(f_m(x,t) - v(t)) - f_m(x,t). \quad (3.10)$$

Then, by substituting (3.10) into (3.7), the following sliding mode dynamics can be obtained:

$$\dot{x}(t) = \sum_{i=1}^{r} h_i(\theta(t))\Big[(A_{i,m} + (\bar{B}_m + \Delta\bar{B}_m)K_{i,m}$$

$$+ \mathcal{I}_m\Delta A_{i,m})x(t) - \Delta B_m(f_m(x,t) - v(t))\Big], \tag{3.11}$$

where $\mathcal{I}_m = I - (\bar{B}_m + \Delta\bar{B}_m)(G\bar{B}_m)^{-1}G$, $\Delta B_m = (\bar{B}_m + \Delta\bar{B}_m)(G\bar{B}_m)^{-1}G\Delta\bar{B}_m$.

The compensator is designed as

$$v(t) = -(\beta_m\|x(t)\| + \varepsilon)sgn(\bar{s}(t)), \tag{3.12}$$

where $\bar{s}(t) = \Delta\bar{B}_m^T P_m x(t)$, with P_m being determined in the following theorem, and $\varepsilon > 0$ is a small constant.

Remark 3.3

The designed sliding surface function is unique, which brings much benefits when compared with other works. Such as in Ref. [27], the input matrices were assumed to have the full column rank; if not, the proposed SMC strategy was unable to tackle the T-S fuzzy models with uncertain parameters. A novel dynamic sliding mode approach was proposed in Refs. [24,25], with which the method in Refs. [33–35] shared the same characteristics. Unfortunately, the designed SMC strategy based on its sliding surface lacked robustness to system uncertainties during the sliding motion phase. In Ref. [28], Lie bracket was introduced to tackle different input matrices, while the method would be very complicated if semi-Markovian switching was taken into consideration of the T-S fuzzy model. Also, the introduced $v(t)$ in (3.12) could eliminate the effect of $\Delta\bar{B}_m f_m(x,t)$, which could ensure good property of the sliding motion.

Remark 3.4

Note that in some works, such as in Ref. [29], individual integral sliding surfaces and controllers were designed in different operating regions of the T-S fuzzy system, which was called piece-wise ISMC. However, the proposed method may be difficult to implement in practice due to its high complexity. In addition, here are some other concerns. First, the continuity of sliding surface functions from one submodel to another is questionable; if not, potential-state jumps will occur. Second, it may cause the following case: the system state has reached onto one of the individual sliding surfaces and converged to the stability performance, but the state is again driven onto another sliding surface with the former stability being destroyed via the piece-wise strategy. More importantly, these endless jumps between sliding surfaces plus chattering effect of sliding mode controllers may be destructive to the system. All these three aspects degenerate the system performance. However, the sliding surface function proposed in Ref. (3.8) is time-continuous and common for all submodels, which can avoid the aforementioned problems.

3.3.2 STOCHASTIC STABILITY ANALYSIS

In this section, a novel stochastic stability condition for the sliding mode dynamics (3.11) is investigated based on generally uncertain TRs.

Theorem 3.1

The sliding mode dynamics (3.11) is robustly stochastically stable, if there exist positive-definite matrix $P_m > 0$ and scalars $\epsilon_{lm} > 0 (l = 1, 2)$ such that the following condition holds for all $m \in S$

$$
\begin{bmatrix}
\Psi_{i,m} & P_m & P_m V_m \\
* & -\epsilon_{1m} \lambda_{\mathcal{I}}^{-1} I & 0 \\
* & * & -\epsilon_{2m} I
\end{bmatrix} < 0,
\tag{3.13}
$$

where

$$
\Psi_{i,m} = \mathrm{He}\{P_m(A_{i,m} + \bar{B}_m K_{i,m})\} + \epsilon_{1m} H_{i,m}^T H_{i,m}
$$

$$
+ \epsilon_{2m} K_{i,m}^T W^T W K_{i,m} + \sum_{n=1}^{s} \pi_{mn}(h) P_m,
$$

$$
\lambda_{\mathcal{I}} = \lambda_{\max}\{\mathcal{I}_m E_{i,m} E_{i,m}^T \mathcal{I}_m^T\},
$$

and in which $K_{i,m}$ is selected such that $A_{i,m} + \bar{B}_m K_{i,m}$ is Hurwitz.

Proof: Consider the following Lyapunov functional of the form:

$$
V(x(t), r_t) = x^T(t) P(r_t) x(t).
\tag{3.14}
$$

Then, according to Definition 1.5, we have

$$
\mathcal{L}V(x(t), m) = \lim_{\delta \to 0} \frac{1}{\delta} \left[\sum_{n=1, n \neq m}^{s} \Pr\{r_{t+\delta} = n \mid r_t = m\} x_\delta^T P_n x_\delta \right.
$$

$$
\left. + \Pr\{r_{t+\delta} = n \mid r_t = m\} x_\delta^T P_m x_\delta - x^T(t) P_m x(t) \right],
$$

where $x_\delta \triangleq x(t+\delta)$. For a general distribution of the sojourn time without memoryless property, that is, $\Pr\{r_{t+\delta} = n \mid r_t = m\} \neq \Pr\{r_\delta = n \mid r_0 = m\}$, by the conditional probability formula, we have

$$
\mathcal{L}V(x(t),m) = \lim_{\delta \to 0} \frac{1}{\delta} \left[\sum_{n=1,n\neq m}^{s} \frac{q_{mn}(G_m(h+\delta)-G_m(t))}{1-G_m(h)} x_\delta^T P_n x_\delta \right.
$$

$$
\left. + \frac{1-G_m(h+\delta)}{1-G_m(h)} x_\delta^T P_m x_\delta - x^T(t)P_m x(t) \right]
$$

$$
= \lim_{\delta \to 0} \frac{1}{\delta} \left[\sum_{n=1,n\neq m}^{s} \frac{q_{mn}(G_m(h+\delta)-G_m(h))}{1-G_m(h)} x_\delta^T P_n x_\delta \right.
$$

$$
+ \frac{1-G_m(h+\delta)}{1-G_m(h)} [x_\delta^T - x^T(t)]P_m x_\delta
$$

$$
+ \frac{1-G_m(h+\delta)}{1-G_m(h)} x^T(t)P_m^T [x_\delta - x(t)]
$$

$$
\left. - \frac{G_m(h+\delta)-G_m(h)}{1-G_m(h)} x^T(t)P_m x(t) \right],
$$

where $G_m(h)$ is the cumulative distribution function of the sojourn time when the system stays in mode m, and q_{mn} is the probability intensity from mode m to mode n. On the other hand, we have that

$$
\lim_{\delta \to 0} \frac{(G_m(h+\delta)-G_m(h))}{(1-G_m(h))\delta}
$$

$$
= \frac{1}{1-G_m(h)} \lim_{\delta \to 0} \frac{(G_m(h+\delta)-G_m(h))}{\delta} = \pi_m(h),
$$

where $\pi_m(h)$ is the TR of the system jumping from mode m. Now, as in Ref. [36], define $\pi_{mn}(h) = \pi_m(h)q_{mn}$ for $n \neq m$ and $\pi_{mm}(h) = -\sum_{n\neq m} \pi_{mn}(h)$. So, the above equality is reduced to

$$
\mathcal{L}V(x(t),m) = 2x^T(t)P_m \dot{x}(t) + x^T(t) \sum_{n=1}^{s} \pi_{mn}(h)P_n x(t)
$$

$$
= 2x^T(t)P_m \sum_{i=1}^{r} h_i(\theta(t)) \left[(A_{i,m} + (\bar{B}_m + \Delta\bar{B}_m)K_{i,m} \right.
$$

$$
\left. + \mathcal{I}_{A_m}\Delta A_{i,m})x(t) - \Delta B_m(f_m(x,t)-v(t)) \right]
$$

$$
+ x^T(t) \sum_{n=1}^{s} \pi_{mn}(h)P_n x(t) \tag{3.15}
$$

On the other hand, it holds that

$$2x^T(t)P_m\Delta\bar{B}_mK_{i,m}x(t) \leq \epsilon_{1m}^{-1}x^T(t)P_mV_mV_m^TP_mx(t)$$

$$+\epsilon_{1m}x^T(t)K_{i,m}^TW^TWK_{i,m}x(t), \qquad (3.16)$$

$$2x^T(t)P_m\mathcal{I}_{A_m}\Delta A_{i,m}x(t) \leq \epsilon_{2m}^{-1}x^T(t)P_m\mathcal{I}_{A_m}E_{i,m}E_{i,m}^T\mathcal{I}_{A_m}^TP_mx(t)$$

$$+\epsilon_{2m}x^T(t)H_{i,m}^TH_{i,m}x(t)$$

$$\leq \epsilon_{2m}^{-1}\lambda_{\mathcal{I}}x^T(t)P_mP_mx(t) + \epsilon_{2m}x^T(t)H_{i,m}^TH_{i,m}x(t), \qquad (3.17)$$

Combining the designed $v(t)$ in (3.12), it holds

$$-2x^T(t)P_m\Delta\mathcal{B}_m(f_m(x,t) - v(t)) \leq -2\varepsilon\,||\,\bar{s}(t)\,|| \leq 0. \qquad (3.18)$$

Overall, it follows that $-2x^T(t)P_m\Delta\mathcal{B}_m(f_m(x,t) - v(t)) \leq -2\varepsilon\,||\,\bar{s}(t)\,|| \leq 0$.

$$\mathcal{L}V(x(t),m) \leq \sum_{i=1}^{r}h_i(\theta(t))x^T(t)\hat{\Gamma}_{i,m}x(t), \qquad (3.19)$$

where $\hat{\Gamma}_{i,m} = \Gamma_{i,m} + \sum_{n=1}^{s}\pi_{mn}(h)P_n$ with

$$\Gamma_{i,m} = He\{P_m(A_{i,m} + \bar{B}_mK_{i,m})\} + \epsilon_{1m}^{-1}P_mV_mV_m^TP_m$$

$$+ \epsilon_{1m}K_{i,m}^TW^TWK_{i,m} + \epsilon_{2m}^{-1}\lambda_{\mathcal{I}}P_mP_m + \epsilon_{2m}H_{i,m}^TH_{i,m}.$$

By Schur complement, it is known from the form of $\hat{\Gamma}_{i,m}$ that the condition (3.13) is equivalent to $\hat{\Gamma}_{i,m} < 0$. Hence, we have

$$\mathcal{L}V(x(t),m) < 0 \text{ for } x(t) \neq 0. \qquad (3.20)$$

Then, it is easy to obtain

$$\lim_{t\to+\infty}\mathbf{E}\left\{\int_0^t ||\,x(s)\,||^2\,ds\,|\,x_0,r_0\right\} < +\infty. \qquad (3.21)$$

Thus, it can be seen from Definition 2.1 that the sliding mode dynamics (3.11) is stochastically stable. Apparently, the stability in the form (3.21) means $x(t)$ converges to 0 with the probability of 1. This completes the proof.

Theorem 3.1 gives the condition under which the sliding mode dynamics (3.11) is stochastically stable, but note that these inequalities are not solvable in MATLAB environment because of the time-varying TRs $\pi_{mn}(h)$. So feasible strict LMI conditions are proposed in the following.

In practice, $\pi_{mn}(h)$ is an uncertain variable, which is not easy to detect in the jumping process. So it is assumed that $\pi_{mn}(h)$ satisfies the following two cases: one is that $\pi_{mn}(h)$ is completely unknown; the other is that $\pi_{mn}(h)$ is not exactly known but upper- and lower-bounded. For instance, $\pi_{mn}(h) \in [\underline{\pi}_{mn}, \bar{\pi}_{mn}]$, in which $\underline{\pi}_{mn}$ and $\bar{\pi}_{mn}$ are the known real constants representing the lower and upper bounds of $\pi_{mn}(h)$, respectively. In view of this, it further denotes $\pi_{mn}(h) \triangleq \pi_{mn} + \Delta\pi_{mn}(h)$, in which $\pi_{mn} = \frac{1}{2}(\underline{\pi}_{mn} + \bar{\pi}_{mn})$ and $|\Delta\pi_{mn}(h)| \leq \lambda_{mn}$ with $\lambda_{mn} = \frac{1}{2}(\bar{\pi}_{mn} - \underline{\pi}_{mn})$. Then, the TR matrix with three jumping modes may be described as

$$\Pi = \begin{bmatrix} \pi_{11} + \Delta\pi_{11}(h) & ? & \pi_{13} + \Delta\pi_{13}(h) \\ ? & ? & \pi_{23} + \Delta\pi_{23}(h) \\ ? & \pi_{32} + \Delta\pi_{32}(h) & \nabla_{33} \end{bmatrix}, \qquad (3.22)$$

where "?" represents the description of unknown TRs. For brevity, $\forall\, m \in S$, let $I_m = I_{m,k} \cup I_{m,uk}$, where

$$I_{m,k} \triangleq \{n : \pi_{mn} \text{ can be determined for } n \in \mathcal{S}\},$$

$$I_{m,uk} \triangleq \{n : \pi_{mn} \text{ is not known for } n \in \mathcal{S}\}.$$

Here, it is assumed that both $I_{m,k} \neq \varnothing$ and $I_{m,uk} \neq \varnothing$. Thus, we can denote the following set:

$$I_{m,k} \triangleq \{k_{m,1}, k_{m,2}, \ldots, k_{m,o}\} \quad 1 \leq o < s,$$

where $k_{m,s}(s \in \{1, 2, \ldots, o\})$ represents the index of s-th element in the m-th row of the TR matrix Π.

Theorem 3.2

The sliding mode dynamics (3.11) is robustly stochastically stable, if there exist positive-definite matrix $P_m > 0$, $T_{mn} > 0$, $Z_{mn} > 0$, and scalars $\epsilon_{lm} > 0 (l = 1, 2)$ such that the following conditions hold for all $m \in S$

If $m \in I_{m,k}$, $\forall l \in I_{m,uk}$, $I_{m,k} \triangleq \{k_{m,1}, k_{m,2}, \ldots, k_{m,o1}\}$,

$$\begin{bmatrix} \mathcal{A}_{i,m}^{11} & P_m & P_m V_m & \mathcal{A}_m^{12} \\ * & -\epsilon_{1m}\lambda_I^{-1}I & 0 & 0 \\ * & * & -\epsilon_{2m}I & 0 \\ * & * & * & \mathcal{A}_m^{13} \end{bmatrix} < 0, \qquad (3.23)$$

If $m \in I_{m,uk}$, $\forall l \in I_{m,uk}$, $I_{m,k} \triangleq \{k_{m,1}, k_{m,2}, \ldots, k_{m,o2}\}$, $l \neq m$,

$$P_m - P_l \geq 0, \qquad (3.24a)$$

$$
\left[
\begin{array}{cccc}
\mathcal{A}_{i,m}^{21} & P_m & P_m V_m & \mathcal{A}_m^{22} \\
* & -\epsilon_{1m}\lambda_I^{-1}I & 0 & 0 \\
* & * & -\epsilon_{2m}I & 0 \\
* & * & * & \mathcal{A}_m^{23}
\end{array}
\right] < 0, \qquad (3.24b)
$$

where

$$
\mathcal{A}_{i,m}^{11} = \mathrm{He}\{P_m(A_{i,m} + \bar{B}_m K_{i,m})\} + \epsilon_{1m} H_{i,m}^T H_{i,m}
$$

$$
+ \sum_{n \in I_{m,k}} \left[\frac{(\lambda_{mn})^2}{4} T_{mn} + \pi_{mn}(P_n - P_l) + \epsilon_{2m} K_{i,m}^T W^T W K_{i,m} \right]
$$

$$
\mathcal{A}_m^{12} = [(P_{k_{m,1}} - P_l) \ \cdots \ (P_{k_{m,o1}} - P_l)]
$$

$$
\mathcal{A}_m^{13} = [-T_{mk_{m,1}} \ \cdots \ -T_{mk_{m,o1}}],
$$

$$
\mathcal{A}_{i,m}^{21} = \mathrm{He}\{P_m(A_{i,m} + \bar{B}_m K_{i,m})\} + \epsilon_{1m} H_{i,m}^T H_{i,m}
$$

$$
+ \sum_{n \in I_{m,k}} \left[\frac{(\lambda_{mn})^2}{4} Z_{mn} + \pi_{mn}(P_n - P_l) + \epsilon_{2m} K_{i,m}^T W^T W K_{i,m} \right]
$$

$$
\mathcal{A}_m^{22} = [(P_{k_{m,1}} - P_l) \ \cdots \ (P_{k_{m,o2}} - P_l)],
$$

$$
\mathcal{A}_m^{23} = [-Z_{mk_{m,1}} \ \cdots \ -Z_{mk_{m,o2}}].
$$

Proof: Case I: $m \in I_{m,k}$

First, denote $\lambda_{m,k} \triangleq \sum_{n \in I_{m,k}} \pi_{mn}(h)$. Since $I_{m,uk} \neq \varnothing$, it holds that $\lambda_{m,k} < 0$. Notice that $\sum_{n=1}^{s} \pi_{mn}(h) P_n$ can be represented as

$$
\sum_{n=1}^{s} \pi_{mn}(h) P_n = \left(\sum_{n \in I_{m,k}} + \sum n \in I_{m,uk} \right) \pi_{mn}(h) P_n
$$

$$
= \sum_{n \in I_{m,k}} \pi_{mn}(h) P_n - \lambda_{m,k} \sum_{n \in I_{m,uk}} \frac{\pi_{mn}(h)}{-\lambda_{m,k}} P_n. \qquad (3.25)
$$

It is obvious that $0 \le \pi_{mn}(h) / -\lambda_{m,k} \le 1 \ (n \in I_{m,uk})$ and $\sum_{n \in I_{m,uk}} \frac{\pi_{mn}(h)}{-\lambda_{m,k}} = 1$. So for $\forall l \in I_{m,uk}$, there is

$$\hat{\Gamma}_{i,m} = \sum_{n \in I_{m,uk}} \frac{\pi_{mn}(h)}{-\lambda_{m,k}} \left[\Gamma_{i,m} + \sum_{n \in I_{m,k}} \pi_{mn}(h)(P_n - P_l) \right]. \tag{3.26}$$

Therefore, for $0 \le \pi_{mn}(h) \le -\lambda_{m,k}$, $\hat{\Gamma}_{i,m} < 0$ is equivalent to

$$\Gamma_{i,m} + \sum_{n \in I_{m,k}} \pi_{mn}(h)(P_n - P_l) < 0. \tag{3.27}$$

In formula (3.27), it is true that

$$\sum_{n \in I_{m,k}} \pi_{mn}(h)(P_n - P_l) = \sum_{n \in I_{m,k}} \pi_{mn}(P_n - P_l) + \sum_{n \in I_{m,k}} \Delta\pi_{mn}(h)(P_n - P_l). \tag{3.28}$$

Then by virtue of Lemma 1.4 and for any $T_{mn} > 0$, it follows

$$\sum_{n \in I_{m,k}} \Delta\pi_{mn}(h)(P_n - P_l) = \sum_{n \in I_{m,k}} \left[\frac{1}{2} \Delta\pi_{mn}(h)((P_n - P_l) + (P_n - P_l)) \right]$$

$$\le \sum_{n \in I_{m,k}} \left[\frac{(\lambda_{mn})^2}{4} T_{mn} + (P_n - P_l)(T_{mn})^{-1}(P_n - P_l)^T \right]. \tag{3.29}$$

From (3.25)–(3.29), it can be concluded that (3.23) guarantees $\hat{\Gamma}_{i,m} < 0$ by applying Schur complement when $m \in I_{m,k}$.

Case II: $m \in I_{m,uk}$.

Similarly, denote $\lambda_{m,k} \triangleq \sum_{n \in I_{m,k}} \pi_{mn}(h)$. Since $I_{m,k} \ne \varnothing$, it holds that $\lambda_{m,k} > 0$. Now, $\sum_{n=1}^{s} \pi_{mn}(h)P_n$ can be represented as

$$\sum_{n=1}^{s} \pi_{mn}(h)P_n = \sum_{n \in I_{m,k}} \pi_{mn}(h)P_n + \pi_{mm}(h)P_m + \sum_{n \in I_{m,uk},n \ne m} \pi_{mn}(h)P_n$$

$$= \sum_{n \in I_{m,k}} \pi_{mn}(h)P_n + \pi_{mm}(h)P_m$$

$$- (\pi_{mm}(h) + \lambda_{m,k}) \sum_{n \in I_{m,uk},j \ne m} \frac{\pi_{mn}(h)P_n}{-\pi_{mm}(h) - \lambda_{m,k}}, \tag{3.30}$$

and it is obvious that $0 \le \pi_{mn}(h)/-\pi_{mm}(h) - \lambda_{m,k} \le 1 \ (n \in I_{m,uk})$ and $\sum_{n \in I_{m,uk},n \ne m} \frac{\pi_{mn}(h)}{-\pi_{mm}(h) - \lambda_{m,k}} = 1$. So for $\forall l \in I_{m,uk}, l \ne i$,

$$\hat{\Gamma}_{i,m} = \sum_{n\in I_{m,uk},n\neq m} \frac{\pi_{mn}(h)}{-\pi_{mm}(h)-\lambda_{m,k}}\left[\Gamma_{i,m}+\pi_{mm}(h)(P_m-P_l)\right.$$

$$\left.+\sum_{n\in I_{m,k}}\pi_{mn}(h)(P_n-P_l)\right] \tag{3.31}$$

Therefore, for $0\leq\pi_{mn}(h)\leq-\pi_{mm}(h)-\lambda_{m,k}$, $\hat{\Gamma}_{i,m}<0$ is equivalent to

$$\Gamma_{i,m}+\pi_{mm}(h)(P_m-P_l)+\sum_{n\in I_{m,k}}\pi_{mn}(h)(P_n-P_l)<0. \tag{3.32}$$

Since $\pi_{mm}(h)<0$, (3.32) holds if we have

$$\begin{cases} P_m-P_l\geq 0, \\ \Gamma_{i,m}+\sum_{n\in I_{m,k}}\pi_{mn}(h)(P_n-P_l)<0. \end{cases} \tag{3.33}$$

Also, as in (3.28) and (3.29), for any $Z_{mn}>0$, we have

$$\sum_{n\in I_{m,k}}\pi_{mn}(h)(P_n-P_l)\leq\sum_{n\in I_{m,k}}\pi_{mn}(P_n-P_l)$$

$$+\sum_{n\in I_{m,k}}\left[\frac{(\lambda_{mn})^2}{4}Z_{mn}+(P_n-P_l)(Z_{mn})^{-1}(P_n-P_l)^T\right]. \tag{3.34}$$

From (3.30)–(3.34), we know that (3.24a) and (3.24b) guarantee $\hat{\Gamma}_{i,m}<0$ by applying Schur complement when $m\in I_{m,uk}$. In summary, the sliding mode dynamics (3.13) is stochastically stable with generally uncertain TRs from the above analysis. This completes the proof.

As has been discussed in many papers, when $B_{i,m}\equiv B_{j,m}(i,j=1,2,\ldots,r)$ hold, $\Delta\bar{B}_m=0$. In this case, we also have the following corollary.

Corollary 3.1

The sliding mode dynamics (3.11) is robustly stochastically stable if there exist positive-definite matrix $P_m>0$, $T_{mn}>0$, $Z_{mn}>0$, and scalars $\epsilon_m>0$ such that the following conditions hold for all $m\in S$

If $m\in I_{m,k}$, $\forall l\in I_{m,uk}$, $I_{m,k}\triangleq\{k_{m,1},k_{m,2},\ldots,k_{m,ol}\}$,

$$\begin{bmatrix} \mathcal{A}_{i,m}^{11} & P_m\mathcal{I}'m & \mathcal{A}_m^{12} \\ * & -\epsilon_m I & 0 \\ * & * & \mathcal{A}_m^{13} \end{bmatrix}<0, \tag{3.35}$$

If $m \in I_{m,uk}$, $\forall l \in I_{m,uk}$, $I_{m,k} \triangleq \{k_{m,1}, k_{m,2}, \ldots, k_{m,o2}\}$, $l \neq m$,

$$P_m - P_l \geq 0, \tag{3.36a}$$

$$\begin{bmatrix} \mathcal{A}_{i,m}^{21} & P_m \mathcal{I}_m' & \mathcal{A}_m^{22} \\ * & -\epsilon_m I & 0 \\ * & * & \mathcal{A}_m^{23} \end{bmatrix} < 0, \tag{3.36b}$$

where $\mathcal{I}_m' = I - B_m (GB_m)^{-1} G$, and

$$\mathcal{A}_{i,m}^{11} = He\{P_m(A_{i,m} + \bar{B}_m K_{i,m})\} + \epsilon_m H_{i,m}^T H_{i,m}$$

$$+ \sum_{n \in I_{m,k}} \left[\frac{(\lambda_{mn})^2}{4} T_{mn} + \pi_{mn}(P_n - P_l) \right],$$

$$\mathcal{A}_{i,m}^{21} = He\{P_m(A_{i,m} + \bar{B}_m K_{i,m})\} + \epsilon_m H_{i,m}^T H_{i,m}$$

$$+ \sum_{n \in I_{m,k}} \left[\frac{(\lambda_{mn})^2}{4} Z_{mn} + \pi_{mn}(P_n - P_l) \right].$$

The other notations are defined in Theorem 3.2.

Remark 3.5

When the input matrices are plant-rule-independent, we can see the designed sliding surface function can be simplified, and no compensator inputs are needed to attenuate the unknown nonlinear disturbances. In this case, the obtained LMI conditions are easy to be checked and the designed T-S fuzzy SMC laws are more easily to be implemented.

3.3.4 REACHABILITY OF SLIDING SURFACE

This part deals with the problem of reachability of the sliding surface $s(t) = 0$; it will be confirmed that with the controller designed in (3.37) and Theorem 3.2 with feasible solutions, $s(t) = 0$ will be reached in a finite-time interval.

Theorem 3.3

Consider the nonlinear semi-Markovian jump T-S fuzzy system (3.1). Suppose the fuzzy sliding surface function is chosen in (3.8) and the conditions in Theorem 3.2 have feasible solutions. Then, the trajectories of the controlled system (3.7) will be driven onto the specified sliding surface $s(t) = 0$ in finite time by the fuzzy SMC law synthesized as follows:

$$u(t) = \sum_{i=1}^{r} h_i(\theta(t))K_{i,m}x(t) + (GB_m)^{-1}G\Delta\bar{B}_m v(t)$$

$$- (GB_m)^{-1}(\rho(t) + \sigma)sgn(s(t)), \qquad (3.37)$$

where $\rho(t) = \sum_{i=1}^{r} h_i(\theta(t)) \| GE_{i,m} \| \|\| H_{i,m}x(t) \| + \rho_m \| G(\bar{B}_m + \Delta\bar{B}_m) \| \|\| x(t) \|$, and σ is a small positive tuning scalar.

Proof: Choose the following Lyapunov functional:

$$V(t) = \frac{1}{2}s^T(t)s(t). \qquad (3.38)$$

Then, the infinitesimal generator of Lyapunov function $V(t)$ along the trajectory of sliding mode dynamics is given as follows:

$$\mathcal{L}V(t) = s^T(t)\dot{s}(t)$$

$$= s^T(t)\sum_{i}^{r} h_i(\theta(t))G\left[\Delta A_{i,m}x(t) + \bar{B}_m(u(t) + f_m(x,t))\right.$$

$$\left. + \Delta\bar{B}_m(f_m(x,t) - v(t)) - \bar{B}_m K_{i,m}x(t)\right].$$

$$\leq -s^T(t)G(\sum_{i}^{r} h_i(\theta(t))\bar{B}_m K_{i,m}x(t) + \Delta\bar{B}_m v(t))$$

$$+ s^T(t)G\bar{B}_m u(t) + | s(t) | \sum_{i}^{r} h_i(\theta(t)) \| GE_{i,m} \| \|\| H_{i,m}x(t) \|$$

$$+ | s(t) | \| G(\bar{B}_m + \Delta\bar{B}_m)f_m(x,t) \| \qquad (3.39)$$

By substituting (3.37) into (3.39), we have

$$\mathcal{L}V(t) \leq -\sigma \| s(t) \| \leq -\sqrt{2}\sigma V^{\frac{1}{2}}(t) < 0, \text{ for } s(t) \neq 0. \qquad (3.40)$$

Based on the variable separation principle, it can be deduced from (3.40) that there exists an instant $t^* = \sqrt{2V(0)}/\sigma$ such that $V(t^*) = 0$; consequently, $s(t) = 0$ for all $t \geq t^*$. Therefore, the reachability of the sliding surface $s(t) = 0$ can be ensured almost in finite time. This completes the proof.

Similarly, we have ISMC strategy for the T-S system (3.1) where the input matrices are plant-rule-independent.

Corollary 3.2

Consider the nonlinear semi-Markovian jump T-S fuzzy system (3.1). Suppose the fuzzy sliding surface function is chosen in Ref. (3.8). Then, the trajectories of the controlled system (3.7) will be driven onto the specified sliding surface $s(t) = 0$ in finite time by the fuzzy SMC law synthesized as follows:

$$u(t) = \sum_{i=1}^{r} h_i(\theta(t)) K_{i,m} x(t) - (GB_m)^{-1} (\rho(t) + \sigma) sgn(s(t)), \qquad (3.41)$$

where $\rho(t) = \sum_{i=1}^{r} h_i(\theta(t)) \| GE_{i,m} \| \| H_{i,m} x(t) \| + \rho_m \| x(t) \|$, and σ is a small positive tuning scalar.

Remark 3.6

As we can see from the proof in (3.39) and particularly in (3.40), initial conditions of the system (3.1) and the tuning scalar σ directly determine the reaching time of the sliding surface. And, the driven time is proportional to the values of initial condition and is inversely proportional to the tuning scalar σ. On the other hand, it is noticed that the *sign* function may produce undesired chattering, which is the main disadvantage of SMC method. From the last century, a few methods have been appeared to reduce the chattering effect, such as the "quasi-sliding mode control," "reaching law method," "dynamic sliding mode method," etc.

Remark 3.7

While implementing $\Delta \bar{B}_m$, we can amplify them by the same way in dealing with $\Delta A_{i,m}$ as $\| \Delta \bar{B}_m \| \leq \| V_m \| \| W \|$ if the membership functions are too complicated. Also, in this paper, it is required that all state components are measurable and the weighting functions are completely known in the ISMC design. However, in many practical systems, these pieces of information may be not available because of environmental factors or high cost. Note that in Ref. [38], observer designs for T-S fuzzy models, in which the weighting functions depend on completely or partially unmeasured variables, subject to unknown inputs and disturbances are investigated. Motivated by the idea proposed in their paper, we will consider the observer-based T-S fuzzy ISMC in our further studies.

Now, the implementation of the SMC strategy on the considered nonlinear semi-Markovian jump T-S fuzzy system (3.1) can be formulated as follows: select appropriate gain matrices $K_{i,m}$ first, then check the feasibility of conditions in (3.23) and (3.24). If it is done, then construct the sliding surface based on (3.8) and sliding mode controller (3.37) with compensator $v(t)$ (3.12).

3.4   NUMERICAL EXAMPLE

In this section, we will present a numerical example to demonstrate the effectiveness of the proposed results. Consider a single-link robot arm in Ref. [37], in which the dynamic equation is given by

$$\ddot{\theta}(t) = -\frac{MgL}{J}\sin(\theta(t)) - \frac{D(t)}{J}\dot{\theta}(t) + \frac{1}{J}u(t),$$

where $\theta(t)$ is the angle position of the arm, and $u(t)$ is the control input. M is the mass of the pay load, J is the moment of inertia, g is the acceleration of gravity, L is the length of the arm, and $D(t)$ is the coefficient of viscous friction. The values of parameters g and L are given by $g = 9.81$ and $L = 0.5$. It is assumed that the parameter $D(t) = D_0 = 2$ is time-invariant, and the parameters M and J have three different modes as shown in Table 3.1.

The transition-probability-rate matrix that relates the three operation modes is given as follows:

$$\Pi = \begin{bmatrix} -1.5 + \Delta\pi_{11}(h) & ? & ? \\ ? & ? & 0.8 + \Delta\pi_{23}(h) \\ 0.5 + \Delta\pi_{31}(h) & ? & -1.0 + \Delta\pi_{33}(h) \end{bmatrix}.$$

Let $x_1(t) = \theta(t)$ and $x_2(t) = \dot{\theta}(t)$, then the nonlinear term $\sin(x_1(t))$ can be represented as [37]

$$\sin(x_1(t)) = h_1(x_1(t))x_1(t) + \beta h_2(x_1(t))x_1(t)$$

with $\beta = 0.01 / \pi$, where $h_1(x_1(t))$, $h_2(x_1(t)) \in [0,1]$, and $h_1(x_1(t)) + h_2(x_1(t)) = 1$. By solving the equations, the membership functions $h_1(x_1(t))$ and $h_2(x_1(t))$ are obtained as follows:

$$h_1(x_1(t)) = \begin{cases} \dfrac{\sin(x_1(t)) - \beta x_1(t)}{x_1(t)(1-\beta)} & x_1(t) \neq 0 \\ 1, & x_1(t) = 0 \end{cases}$$

TABLE 3.1
Modes and Values of the Parameters M and J

Mode m	Parameter M	Parameter J
1	1	1
2	1.5	2
3	2	2.5

$$h_2(x_1(t)) = \begin{cases} \dfrac{x_1(t) - \sin(x_1(t))}{x_1(t)(1-\beta)} & x_1(t) \neq 0 \\[2mm] 0, & x_1(t) = 0 \end{cases}$$

It is evident from the abovementioned membership functions that when $x_1(t)$ is about 0 rad, $h_1(x_1(t)) = 1, h_2(x_1(t)) = 0$, and when $x_1(t)$ is about π rad or $-\pi$ rad, $h_1(x_1(t)) = 0, h_2(x_1(t)) = 1$. Thus, when taking system parameter uncertainties into consideration, the state space representation of single-link robot arm can be expressed by the following two-rule T-S fuzzy system:

Plant Rule 1: IF $x_1(t)$ is "about 0 rad,"
THEN

$$\dot{x}(t) = (A_{1,m} + \Delta A_{1,m})x(t) + B_{1,m}(u(t) + f_m(x(t))).$$

Plant Rule 2: IF $x_1(t)$ is "about π rad or $-\pi$ rad,"
THEN

$$\dot{x}(t) = (A_{2,m} + \Delta A_{2,m})x(t) + B_{2,m}(u(t) + f_m(x(t))).$$

where $x(t) = [x_1^T(t) \ x_2^T(t)]^T$, and

$$A_{1,1} = \begin{bmatrix} 0 & 1 \\ -gL & -D_0 \end{bmatrix}, B_{1,1} = B_{2,1} = \begin{bmatrix} 0 \\ 1 \end{bmatrix},$$

$$A_{1,2} = \begin{bmatrix} 0 & 1 \\ -0.75gL & -0.5D_0 \end{bmatrix}, B_{1,2} = B_{2,2} = \begin{bmatrix} 0 \\ 0.5 \end{bmatrix},$$

$$A_{1,3} = \begin{bmatrix} 0 & 1 \\ -0.8gL & -0.4D_0 \end{bmatrix}, B_{1,3} = B_{2,3} = \begin{bmatrix} 0 \\ 0.4 \end{bmatrix},$$

$$A_{2,1} = \begin{bmatrix} 0 & 1 \\ -\beta gL & -D_0 \end{bmatrix}, A_{2,2} = \begin{bmatrix} 0 & 1 \\ -0.75\beta gL & -0.5D_0 \end{bmatrix},$$

$$A_{2,3} = \begin{bmatrix} 0 & 1 \\ -0.8\beta gL & -0.4D_0 \end{bmatrix}, E_{i,1} = \begin{bmatrix} 0 \\ 0.3 \end{bmatrix}, E_{i,2} = \begin{bmatrix} 0.1 \\ 0 \end{bmatrix}, E_{i,3} = \begin{bmatrix} 0.2 \\ 0.1 \end{bmatrix},$$

$$H_{i,1} = [0.1 \ -0.1], H_{i,2} = [-0.1 \ 0.2], H_{i,2} = [0 \ 0.1], (i \in \{1,2\}).$$

It is selected that $\beta_m = 0.1 \ (m = 1,2,3)$. In this model, the input matrices are plant-rule-independent, so we can use the conditions in Corollary 3.1 to check the effectiveness of the proposed SMC theory. Let $G = [0 \ 1]$ such that GB_m is nonsingular, $K_{i,m} = [-2 \ -3] \ (i \in \{1,2,\dots,r\}, m \in S)$ and $\Delta\pi_{mn}(h) \le \lambda_{mn} = |0.1 * \pi_{mn}|$. By solving (30) and (31), we can obtain the following feasible solutions:

$$P_1 = \begin{bmatrix} 5.3941 & 0.7299 \\ 0.7299 & 0.8312 \end{bmatrix}, P_2 = \begin{bmatrix} 5.9056 & 1.0500 \\ 1.0500 & 1.8304 \end{bmatrix},$$

$$P_3 = \begin{bmatrix} 5.7512 & 1.1380 \\ 1.1380 & 1.8093 \end{bmatrix}, T_{1,1} = \begin{bmatrix} 5.7453 & 0.0446 \\ 0.0446 & 5.8212 \end{bmatrix},$$

$$T_{3,1} = \begin{bmatrix} 5.6599 & 0.0451 \\ 0.0451 & 5.7602 \end{bmatrix}, T_{3,3} = \begin{bmatrix} 5.6936 & 0.0028 \\ 0.0028 & 5.6720 \end{bmatrix},$$

$$Z_{2,3} = \begin{bmatrix} 5.6416 & 0.0436 \\ 0.0436 & 5.7478 \end{bmatrix}, \epsilon_1 = 5.6063, \epsilon_2 = 5.5482, \epsilon_3 = 5.7059.$$

Given the initial condition $x(0) = [0.5\pi \ -1]^T$, the adjustable parameters $\sigma = 0.05$ and $f_m(x(t)) = 0.1\sin(x_1(t))$ $(m = 1,2,3)$. Also, in order to reduce the chattering effect of switching signals, $\text{sgn}(s(t))$ is replaced by $\dfrac{s(t)}{\|s(t)\| + 0.01}$. The simulation results are presented in Figures 3.1–3.4. Figure 3.1 plots the membership functions. Figure 3.2 draws the state response of the closed-loop system. Figure 3.3 shows the control input. We can see from these figures that the system achieved better stochastic stability by the designed ISMC laws. Also, we note that the tuning scalar σ directly matters when the sliding surface will be reached. It is known from (3.37) that for the fixed initial condition, the bigger the value of σ is, then faster the sliding surface will be reached. But this may not be true for each individual simulation, because σ only limits the upper bound of the reaching time and the system may run differently under different jumping modes, which is in accordance with the simulation results presented in Figures 3.4 and 3.5, in which the jumping modes follow the same principle in Figure 3.4. In addition, in order to show the superiority of sliding mode that is insensitive to nonlinear perturbations, a simulation in Figure 3.6 is presented via two control methods with the same gain matrices as before. As we can see from this figure, in which the green lines represent the response by the ISMC method and the pink lines represent the response by the control method proposed in Ref. [37], the green lines go smoother than the pink ones. Also, it can be derived from the form of sliding mode dynamics (3.11), the stronger the uncertainties are imposed on the system, the better superiority of ISMC will be shown.

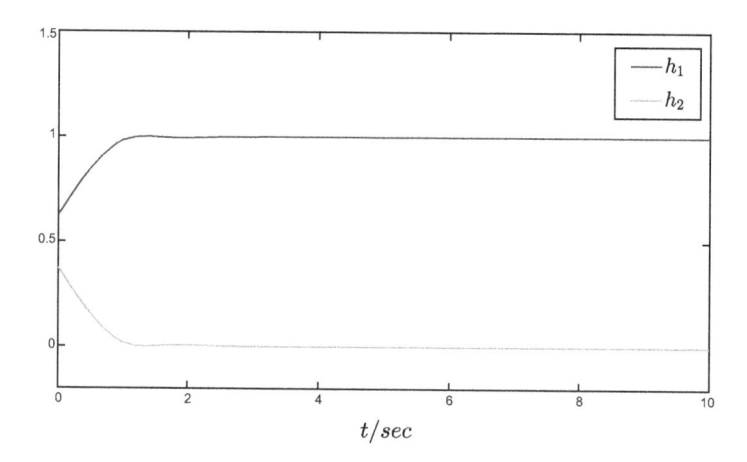

FIGURE 3.1 Membership functions.

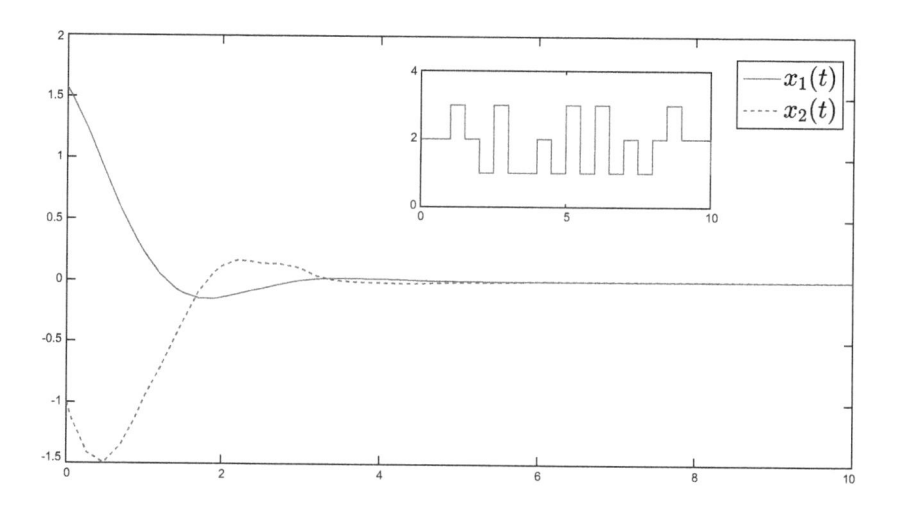

FIGURE 3.2 Response of system state $x(t)$.

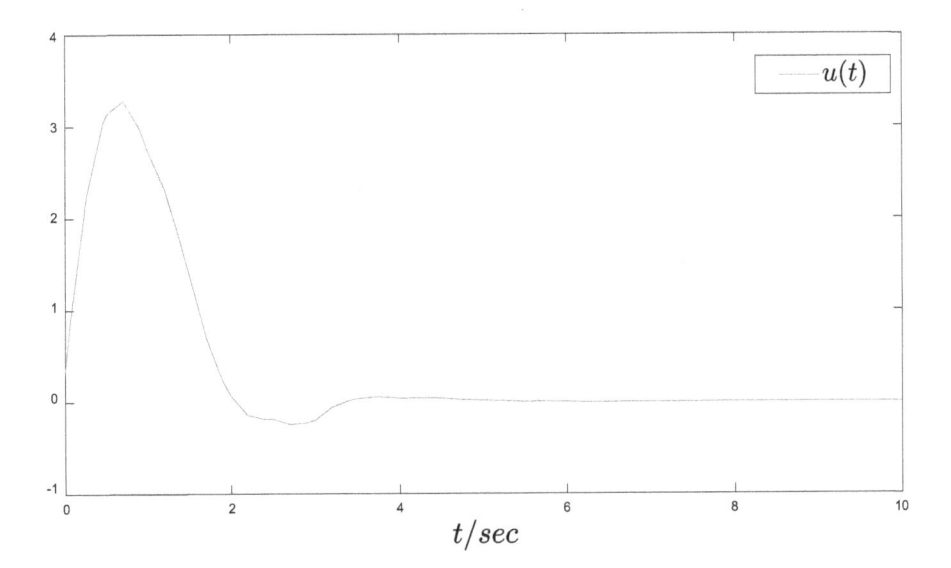

FIGURE 3.3 Control input $u(t)$.

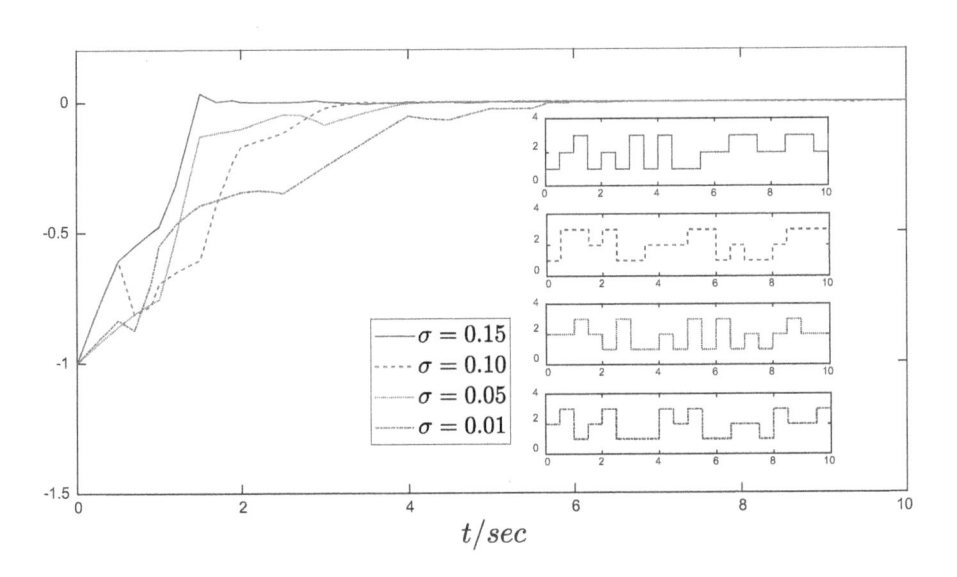

FIGURE 3.4 Sliding surface function with different tuning scalar σ.

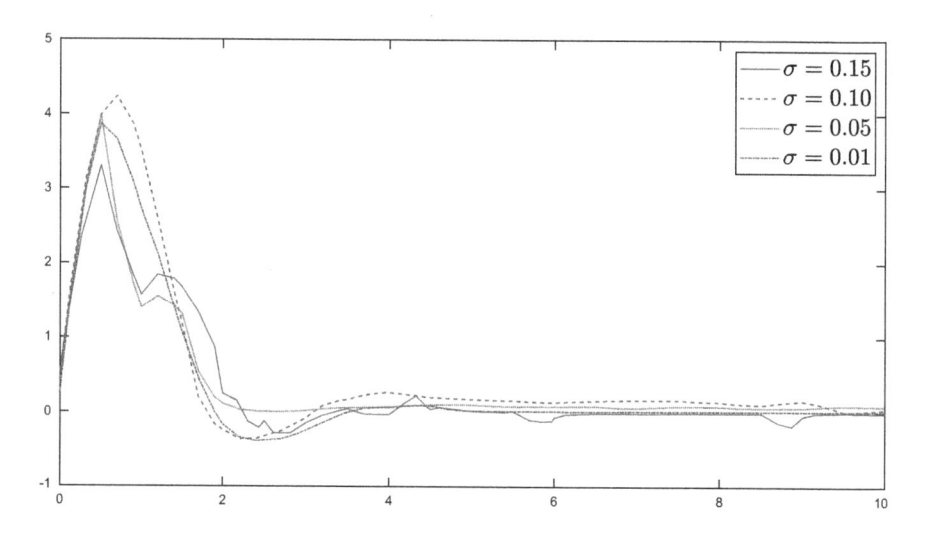

FIGURE 3.5   Control input with different tuning scalar σ.

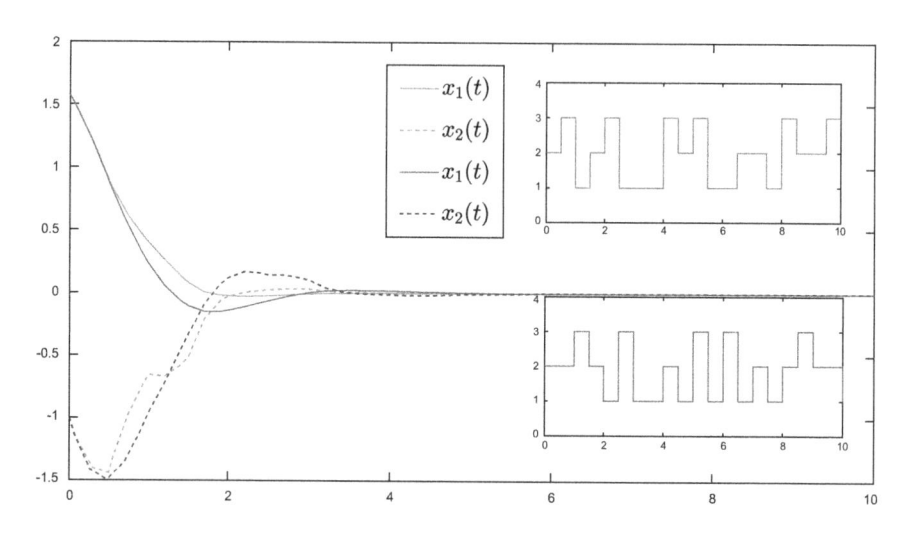

FIGURE 3.6   State response of $x(t)$ via ISMC (green lines) and the method proposed in Ref. [37] (pink lines).

3.5 CONCLUSION

In this chapter, a novel fuzzy integral SMC strategy has been proposed for a class of nonlinear semi-Markovian jump T-S fuzzy systems. A new fuzzy integral switching surface function has been put forward based on different input matrices, and feasible LMI conditions that ensure the stochastic stability of the sliding mode dynamics with generally uncertain TRs has been derived. Then, fuzzy SMC laws have been synthesized to drive the system trajectories onto the proposed fuzzy switching surface. Finally, a practical example has been offered to verify the superiority and effectiveness of the achieved fuzzy ISMC scheme. As mentioned before, for the complexity of system structure, the study of adaptive and observer-based SMC of T-S fuzzy systems with uncertain premise variables will be very attractive in the following.

REFERENCES

[1] T. Takagi and M. Sugeno, Fuzzy identification of systems and its application to modeling and control, *IEEE Transactions on Systems, Man, and Cybernetics, Part B, Cybernetics*, Vol. 15, No. 1, pp. 116–132, 1985.

[2] L. Wu, X. Su, P. Shi, and J. Qiu, A new approach to stability analysis and stabilization of discrete-time T-S fuzzy time-varying delay systems, *IEEE Transactions on Systems, Man, and Cybernetics, Part B, Cybernetics*, Vol. 41, No. 1, pp. 273–286, 2011.

[3] E. Kim, and H. Lee, New approaches to relaxed quadratic stability condition of fuzzy control systems, *IEEE Transactions on Fuzzy Systems*, Vol. 8, No. 5, pp. 523–534, 2000.

[4] L. Zhao, H. Gao, and H. R. Karimi, Robust stability and stabilization of uncertain T-S fuzzy systems with time-varying delay: An input output approach, *IEEE Transactions on Fuzzy Systems*, Vol. 21, No. 5, pp. 883–897, 2013.

[5] X. Su, P. Shi, L. Wu, and M. V. Basin, Reliable filtering with strict dissipativity for T-S fuzzy time-delay systems, *IEEE Transactions on Cybernetics*, Vol. 44, No. 12, pp. 2470–2483, 2014.

[6] C. N. Bhende, S. Mishra, and S. K. Jain, TS-fuzzy-controlled active power filter for load compensation, *IEEE Transactions on Power Delivery*, Vol. 21, No. 3, pp. 1459–1465, 2006.

[7] J. Qiu, G. Feng, and J. Yang, A new design of delay-dependent robust H1 filtering for discrete-time T-S fuzzy systems with time-varying delay, *IEEE Transactions on Fuzzy Systems*, Vol. 17, No. 5, pp. 1044–1058, 2009.

[8] S. W. Kau, H. J. Lee, C. M. Yang, C. Lee, and L. Hong, Robust H_∞ fuzzy static output feedback control of T-S fuzzy systems with parametric uncertainties, *Fuzzy Sets and Systems*, Vol. 158, No. 2, pp. 135–146, 2007.

[9] J. Qiu, G. Feng, and H. Gao, Static-output-feedback H_∞ control of continuous-time T-S fuzzy affine systems via piecewise Lyapunov functions, *IEEE Transactions on Fuzzy Systems*, Vol. 21, No. 2, pp. 245–261, 2013.

[10] H. Shen, Y. Men, Z. G. Wu, and J. H. Park, Nonfragile H_∞ control for fuzzy Markovian jump systems under fast sampling singular perturbation, *IEEE Transactions on Systems, Man, and Cybernetics: Systems*, Vol. 48, No. 12, pp. 2058–2069, 2017.

[11] B. Jiang, Z. Gao, P. Shi, and Y. Xu, Adaptive fault-tolerant tracking control of near-space vehicle using Takagi-Sugeno fuzzy models, *IEEE Transactions on Fuzzy Systems*, Vol. 18, No. 5, pp. 1000–1007, 2010.

[12] Y. J. Liu, S.C. Tong, and T. S. Li, Observer-based adaptive fuzzy tracking control for a class of uncertain nonlinear MIMO systems, *Fuzzy Sets and Systems*, Vol. 164, No. 1, pp. 25–44, 2011.

[13] J. J. Rubio, USNFIS: Uniform stable neuro fuzzy inference system, *Neurocomputing*, Vol. 262, pp. 57–66, 2017.

[14] L. A. Paramo-Carranza, J. A. Meda-Campaa, J. J. Rubio, R. T. Herrera, A. V. C. Lopez, A. G. Meza and I. Czares-Ramirez, Discrete-time Kalman filter for Takagi-Sugeno fuzzy models, *Evolving Systems*, Vol. 8, No. 3, pp. 211–219, 2017.

[15] A. Grande, T. Hernandez, A. V. Curtidor, L. A. Paramo, R. Tapia, I. O. Cazares and J. A Meda, Analysis of fuzzy observability property for a class of TS fuzzy models, *IEEE Latin America Transactions*, Vol. 15, No. 4, pp. 595–602, 2017.

[16] H. H. Choi, Robust stabilization of uncertain fuzzy systems using variable structure system approach, *IEEE Transactions on Fuzzy Systems*, Vol. 16, No. 3, pp. 715–724, 2008.

[17] J. Zhang, P. Shi, and Y. Xia, Robust adaptive sliding-mode control for fuzzy systems with mismatched uncertainties, *IEEE Transactions on Fuzzy Systems*, Vol. 18, No. 4, pp. 700–711, 2010.

[18] H. Li, J. Wang, and P. Shi, Output-feedback based sliding mode control for fuzzy systems with actuator saturation, *IEEE Transactions on Fuzzy Systems*, Vol. 24, No. 6, pp. 1282–1293, 2016.

[19] H. H. Choi, Adaptive controller design for uncertain fuzzy systems using variable structure control approach, *Automatica*, Vol. 45, No. 11, pp. 2646–2650, 2009.

[20] Z. Xi, G. Feng, T. Hesketh, Piecewise sliding-mode control for T-S fuzzy systems, *IEEE Transactions on Fuzzy Systems*, Vol. 19, No. 4, pp. 707–716, 2011.

[21] S. Wen, T. Huang, X. Yu, and M. Chen, Aperiodic sampled-data sliding-mode control of fuzzy systems with communication delays via the event-triggered method, *IEEE Transactions on Fuzzy Systems*, Vol. 24, No. 5, pp. 1048–1057, 2016.

[22] C. L. Hwang, A novel Takagi-Sugeno-based robust adaptive fuzzy sliding-mode controller, *IEEE Transactions on Fuzzy Systems*, Vol. 12, No. 5, pp. 676–687, 2004.

[23] D. W. C. Ho, and Y. Niu, Robust fuzzy design for nonlinear uncertain stochastic systems via sliding-mode control, *IEEE Transactions on Fuzzy Systems*, Vol. 15, No. 3, pp. 350–358, 2007.

[24] Q. Gao, G. Feng, Z. Xi, and Y. Wang, A new design of robust H_∞ sliding mode control for uncertain stochastic T-S fuzzy time-delay systems, *IEEE Transactions on Cybernetics*, Vol. 44, No. 9, pp. 1556–1566, 2014.

[25] Q. Gao, G. Feng, Z. Xi, Y. Wang, and J. Qiu, Robust H_∞ control of T-S fuzzy time-delay systems via a new sliding-mode scheme, *IEEE Transactions on Fuzzy Systems*, Vol. 22, No. 2, pp. 459–465, 2014.

[26] H. Li, J. Wang, H. K. Lam, and Q. Zhou, Adaptive sliding mode control for interval type-2 fuzzy systems, *IEEE Transactions on Systems, Man, and Cybernetics: Systems*, Vol. 46, No. 12, pp. 1654–1663, 2016.

[27] H. Li, J. Wang, H. Du, and H. R. Karimi, Adaptive sliding mode control for Takagi-Sugeno fuzzy systems and its applications, *IEEE Transactions on Fuzzy Systems*, Vol. 26, No. 2, pp. 531–542, 2018.

[28] Y. Wang, H. Shen, H. R. Karimi, and D. P. Duan, Dissipativity-based fuzzy integral sliding mode control of continuous-time T-S fuzzy systems, *IEEE Transactions on Fuzzy Systems*, Vol. 26, No. 3, pp. 1164–1176, 2018.

[29] Z. Xi, G. Feng, and T. Hesketh, Piecewise integral sliding-mode control for TS fuzzy systems, *IEEE Transactions on Fuzzy Systems*, Vol. 19, No. 1, pp. 65–74, 2011.

[30] Q. Gao, L. Liu, G. Feng, and Y. Wang, Universal fuzzy integral sliding-mode controllers for stochastic nonlinear systems, *IEEE Transactions on Cybernetics*, Vol. 44, No. 12, pp. 2658–2669, 2014.

[31] J. Li, Q. Zhang, X. G. Yan, and S. K. Spurgeon, Robust stabilization of T-S fuzzy stochastic descriptor systems via integral sliding modes, *IEEE Transactions on Cybernetics*, Vol. 48, No. 9, pp. 2736–2749, 2017.

[32] J. Li, Q. Zhang, X. G. Yan, and S. K. Spurgeon, Integral sliding mode control for Markovian jump T-S fuzzy descriptor systems based on the super-twisting algorithm, *IET Control Theory & Applications*, Vol. 11, No. 8, pp. 1134–1143, 2016.

[33] J. J. Rubio, E. Soriano, C. F. Juarez and J. Pacheco, Sliding mode regulator for the perturbations attenuation in two tank plants, *IEEE Access*, Vol. 5, pp. 20504–20511, 2017.

[34] C. A. Ibanez, Stabilization of the PVTOL aircraft based on a sliding mode and a saturation function, *International Journal of Robust and Nonlinear Control*, Vol. 27, No. 5, pp. 843–859, 2017.

[35] J. J. Rubio, Robust feedback linearization for nonlinear processes control, *ISA Transactions*, Vol. 74, pp. 155–164, 2018.

[36] J. Huang, and Y. Shi, H_∞ state-feedback control for semi-markov jump linear systems with time-varying delays, *Journal of Dynamic Systems, Measurement, and Control*, Vol. 135, No. 4, 041012, 2013.

[37] H. N. Wu and K. Y. Cai, Mode-independent robust stabilization for uncertain Markovian jump nonlinear systems via fuzzy control, *IEEE Transactions on Systems, Man, and Cybernetics: Systems*, Vol. 36, No. 3, pp. 509–519, 2005.

[38] M. Chadli and H. R. Karimi, Robust observer design for unknown inputs Takagi-Sugeno models, *IEEE Transactions on Fuzzy Systems*, Vol. 21, No. 1, pp. 158–164, Feb. 2013.

4 Fuzzy Sliding Mode Control for Finite-Time Synthesis of Semi-Markovian Jump Systems

4.1 INTRODUCTION

As is well known, the stability problem is one of the fundamental issues in control community. Conveniently, the results reported on stability was often referred to as Lyapunov stability (LS). One of the typical properties of LS is that the solution of the system will tend to an equilibrium state as time tends to infinity. While many real systems are only concerned with the performance over a fixed or finite-time interval because it is not acceptable for large values of system state in the presence of saturation [1], since it not allows the state to exceed a prescribed value during a finite-time interval for the given bound on the initial condition. More importantly, most of works are only related to stability and performance criteria based on classical stability theory defined over an infinite-time interval. Therefore, Peter Dorato [2] proposed the finite-time stability for analysis. And later, Amato et al. [3] engaged in finite-time boundedness analysis by taking initial conditions and external disturbance into consideration. Recently, focusing on the finite-time stability analysis, the sliding mode control (SMC) approach has been immersed with finite-time control. For example, the finite-time stabilization problem for nonlinear systems by SMC approach was presented in Refs. [4,5], while these studies were based on the exact information of the state variables. And finite-time SMC for unmeasurable state Markovian jump system (MJS) was investigated in Ref. [5]; while as we know, an SMC scheme involves two steps: the sliding motion phase and the reaching phase – but the finite-time boundedness analysis was only carried out during the latter phase in Ref. [6] – the boundedness performance during the reaching phase was neglected, so one question that how to regulate the finite-time boundedness performance of the system during the reaching phase raises? Therefore, the finite-time fuzzy SMC for semi-Markovian jump systems (S-MJSs) is a tough problem to be tackled by considering its unmeasurable state variables, unmeasurable premise variables and generally uncertain time-varying transition rates.

Based on the above analysis, this chapter will concern with the problem of observer-based fuzzy SMC for finite-time synthesis of S-MJSs, in which the premise variables are not measurable. The main issues to be investigated include the following: (1) the observer-based finite-time fuzzy SMC is proposed for S-MJSs with immeasurable premise variables and generally uncertain TRs; (2) a fuzzy sliding mode controller is designed to guarantee that the sliding surface can be reached in finite time before the prescribed time; (3) the finite-time boundedness of the observer-based controlled system is ensured during the reaching phase and sliding motion phase simultaneously; (4) establishing comprehensive feasible conditions to ensure the fuzzy observer system with the error dynamics is finite-time-bounded during the whole phase.

4.2 SYSTEM DESCRIPTION

Consider the following S-MJS by a T-S fuzzy model on the probability space $(\Omega, \mathcal{F}, \mathcal{P})$:
Plant Rule i : **IF** $x_1(t)$ is M_{i1} **and** $x_2(t)$ is M_{i2} **and** $\cdots$ **and** $x_n(t)$ is M_{in}
THEN

$$\begin{cases} \dot{x}(t) = A_i(r_t)x(t) + B_i(r_t)u(t) \\ y(t) = C(r_t)x(t) \\ x(0) = \varphi(0) \end{cases}, \tag{4.1}$$

where $x(t) = [x_1(t) \ x_2(t) \cdots \ x_n(t)]^T \in \mathbb{R}^n$ denotes the state vector; $x_1(t)$, ..., $x_n(t)$ are also selected as premise variables; $M_{ij}(i = 1, 2 \ldots, r; j = 1, 2 \ldots, n)$ are the fuzzy sets. $u(t) \in \mathbb{R}^m$ is the control input; $y(t) \in \mathbb{R}^p$ is the controlled output. $A_i(r_t)$, $B_i(r_t)$ and $C(r_t)$ are the system matrices with appropriate dimensions.

$\{r_t, t \geq 0\}$ is a continuous-time semi-Markovian process, taking discrete values in a finite set $S = \{1, 2, \ldots, s\}$ with transition probabilities governed by:

$$\Pr\{r_{t+h} = n \mid r_t = m\} = \begin{cases} \pi_{mn}(h)h + o(h), & m \neq n, \\ 1 + \pi_{mm}(h)h + o(h), & m = n, \end{cases} \tag{4.2}$$

where $h > 0$ and $\lim_{h \to 0} o(h) / h = 0$, $\pi_{mn}(h) > 0, m \neq n$, is the transition rate from mode m at time t to mode n at time $t + h$, and $\pi_{mm}(h) = -\sum_{n \neq m} \pi_{mn}(h) < 0$ for each $n \in \mathcal{S}$. The generator matrix is defined by $\Pi = [\pi_{mn}(h)]_{s \times s}$. For simplicity, $r(t) = m$ is used for notation in the following.

From a practical point of view, the time-varying TRs in semi-Markov processes are not easy to be acquired directly due to the consideration of the high complexity of system structure. Thus, in this note, more generally uncertain TRs are investigated, and it is considered that the TRs $\pi_{mn}(h)$ satisfy one of the following two conditions:

I. : $\pi_{mn}(h)$ is completely unknown;
II. : $\pi_{mn}(h)$ is not exactly known but upper- and lower-bounded.

For the case (II), for instance, it is assumed that $\pi_{mn}(h) \in [\underline{\pi}_{mn}, \overline{\pi}_{mn}]$, in which $\underline{\pi}_{mn}$ and $\overline{\pi}_{mn}$ are the known real constants representing the lower and upper bounds of $\pi_{mn}(h)$, respectively. In view of this, we further denote $\pi_{mn}(h) \triangleq \pi_{mn} + \Delta\pi_{mn}(h)$, in which

$\pi_{mn} = \frac{1}{2}(\underline{\pi}_{mn} + \bar{\pi}_{mn})$ and $|\Delta\pi_{mn}(h)| \le \lambda_{mn}$ with $\lambda_{mn} = \frac{1}{2}(\bar{\pi}_{mn} - \underline{\pi}_{mn})$. So the TR matrix with three jumping modes may be described as

$$\begin{bmatrix} \pi_{11} + \Delta\pi_{11}(h) & ? & \pi_{13} + \Delta\pi_{13}(h) \\ ? & ? & \pi_{23} + \Delta\pi_{23}(h) \\ ? & \pi_{32} + \Delta\pi_{32}(h) & ? \end{bmatrix}, \qquad (4.3)$$

where "?" represents the description of unknown TRs. For brevity, $\forall\, m \in \mathcal{S}$, let $I_m = I_{m,k} \cup I_{m,uk}$, where

$$I_{m,k} \triangleq \{n : \pi_{mn} \text{ can be determined for } n \in \mathcal{S}\},$$

$$I_{m,uk} \triangleq \{n : \pi_{mn} \text{ is not known for } n \in \mathcal{S}\}.$$

Here, it is assumed that both $I_{m,k} \ne \varnothing$ and $I_{m,uk} \ne \varnothing$. Thus, we can denote the following set:

$$I_{m,k} \triangleq \{k_{m,1}, k_{m,2}, \dots, k_{m,o}\} \quad 1 \le o < s,$$

where $k_{m,s}(s \in \{1, 2, \dots, o\})$ represents the index of s-th element in the m-th row of the TR matrix Π.

Remark 4.1

Here, $I_{m,k} \equiv \varnothing$ and $I_{m,k} \equiv \{1, 2, \dots, s\}$ are not considered, because these two cases represent one totally bad situation and one fully ideal situation, respectively. For the former case, it is not easy to give good results so far, while the latter case has been studied by a few.

By fuzzy blending, that is, adopting a center-average defuzzifier, product-fuzzy inference and a singleton fuzzifier, overall fuzzy model (4.1) is inferred as follows:

$$\begin{cases} \dot{x}(t) = \sum_{i=1}^{r} h_i(x(t))[A_i(r_t)x(t) + B_i(r_t)u(t)] \\ y(t) = C(r_t)x(t) \end{cases}, \qquad (4.4)$$

where $h_i(x(t))$ is the membership function given by $h_i(x(t)) = \dfrac{\Pi_{j=1}^{n}\mu_{ij}(x_j(t))}{\sum\limits_{i=1}^{r}\Pi_{j=1}^{n}\mu_{ij}(x_j(t))}$, with

$\mu_{ij}(x_j(t))$ being the grade of membership of $x_j(t)$ in μ_{ij}. For all $t > 0$, it is satisfied that $h_i(x(t)) \ge 0$ and $\sum\limits_{i=1}^{r} h_i(x(t)) = 1$.

Assumption 4.1

The input matrices satisfy $B_{1,m} = B_{2,m} = \ldots = B_{r,m}$ and have full column rank.

Assumption 4.2

The pairs $(A_{i,m}, B_m)$ are controllable and $(A_{i,m}, C_m)$ are observable.

Consider the premise variables in the system state that are not available. Therefore, the following fuzzy state observer is designed:

Observer Rule i : **IF** $\hat{x}_1(t)$ is M_{i1} **and** $\hat{x}_2(t)$ is M_{i2} **and** $\cdots$ **and** $\hat{x}_n(t)$ is M_{in}
THEN

$$\begin{cases} \dot{\hat{x}}(t) = A_{i,m}\hat{x}(t) + B_m u(t) + L_{i,m}(y(t) - \hat{y}(t)) \\ \hat{y}(t) = C_m \hat{x}(t) \\ \hat{x}(0) = \phi(0) \end{cases} \tag{4.5}$$

where $\hat{x}(t)$ and $\hat{y}(t)$ are the estimation of the state $x(t)$ and output $y(t)$, respectively. $L_{i,m}$ is the observer gain with appropriate dimensions to be determined later.

Then, the fuzzy observer (4.5) is inferred as

$$\begin{cases} \dot{\hat{x}}(t) = \sum_{i=1}^{r} h_i(\hat{x}(t))[A_{i,m}\hat{x}(t) + B_m u(t) + L_{i,m}(y(t) - \hat{y}(t))], \\ \hat{y}(t) = C_m \hat{x}(t) \end{cases} \tag{4.6}$$

Define $\theta(t) \triangleq x(t) - \hat{x}(t)$ as the estimation error. Combining (4.4) and (4.6) yields the error dynamics as follows:

$$\dot{\theta}(t) = \sum_{i=1}^{r} h_i(\hat{x}(t))[(A_{i,m} - L_{i,m}C_m)\theta(t)] + w(t), \tag{4.7}$$

where $w(t) = \sum_{i=1}^{r} (h_i(x(t)) - h_i(\hat{x}(t)))\dot{x}(t)$ is seen as disturbance, and it is assumed that $w(t)$ satisfies

$$\int_0^T w^T(s)w(s)ds \leq d, \quad d \geq 0, \tag{4.8}$$

where d is a known constant.

Remark 4.2

In many other works, the premise variables are assumed to be measurable independent of the designed observer, which limits the practical application of the model. In Ref. [7], robust observer design for unknown inputs in T-S models was proposed with immeasurable premise variables. However, the premise variables are estimated independently, while in many practical problems, the premise variables may depend on the unmeasurable state variables and a fuzzy system describes the widest class of nonlinear systems when the premise variables are the system states [8]. So to have a real sliding mode observer design, the premise variables of the fuzzy observer depending on the estimated state variables by the fuzzy observer make sense.

A system is considered to be finite-time-bounded (FTB) if given initial condition, the state remains within the prescribed limit over a specific finite time for all admissible inputs.

Definition 4.1

Given two scalars $c_1 > 0$ and $c_2 > 0$ with $c_1 < c_2$, a weighting matrix $R > 0$, and time interval $[0, T]$, the fuzzy observer system (4.5) is FTB with respect to $\{c_1, c_2, T, R, d\}$, if for $\forall t \in [0, T]$

$$\hat{x}^T(0)R\hat{x}(0) + \theta^T(0)R\theta(0) \leq c_1 \quad \Rightarrow \quad \mathbf{E}[\hat{x}^T(t)R\hat{x}(t)] \leq c_2. \tag{4.9}$$

Remark 4.3

In view of the above definitions, the concept of finite-time stability is different from that of Lyapunov asymptotic stability, and the general idea of finite-time stability concerns the boundedness of the state over a finite-time interval. On one hand, it is obvious that finite-time stable systems could be Lyapunov stable, whereas a Lyapunov stable system may not be finite-time stable if its state exceeds the prescribed bounds during the transient period. On the other hand, different from the concept used in nonlinear systems that finite-time stability means the state trajectories can reach an equilibrium point within a finite-time interval while still maintaining the classical characteristics of stability, obviously, this is not the situation we considered in this paper, but both of them are important for different practical application.

The purpose of this part can be concluded as investigating finite-time boundedness of the fuzzy observer system (4.6) with the error dynamics (4.7) via sliding mode approach. As we all know, SMC includes two phases: reaching phase and sliding motion phase. Thus, the property (4.9) must be ensured during both phases, and it must also be ensured that when the sliding surface is reached at time T^*, it holds that $T^* \leq T$.

4.3 MAIN RESULTS

4.3.1 FINITE-TIME REACHABILITY OF SLIDING SURFACE IN T^*

Based on the state observer (4.6), we propose the fuzzy integral sliding surface function as:

$$s(t) = G\hat{x}(t) - \int_0^t \sum_{i=1}^r h_i(\hat{x}(s))G(A_{i,m} + B_m K_{i,m})\hat{x}(s)ds, \tag{4.10}$$

where $G \in \mathbb{R}^{m \times n}$ is chosen such that GB_m is nonsingular, and $K_{i,m} \in \mathbb{R}^{m \times n}$ is selected such that $A_{i,m} + B_m K_{i,m}$ is Hurwitz.

In the following, an observer-based sliding mode controller will be designed such that the sliding surface $s(t) = 0$ can be reached before the prescribed time T, and the reaching time $T^* \leq T$ is also ensured. Further, the controller keeps the state remain on the sliding surface thereafter.

Theorem 4.1

Given the designed fuzzy observer system (4.6), the sliding surface function is proposed in (4.10). Then, the sliding surface $s(t) = 0$ can be reached in finite time within T^* by the fuzzy SMC law designed as follows:

$$u(t) = \sum_{i=1}^{r} h_i(\hat{x}(t)) K_{i,m} \hat{x}(t) - (GB_m)^{-1}(\rho(t) + \delta) sgn(s(t)) \tag{4.11}$$

with

$$\rho(t) = \max_{m \in S} \sum_{i=1}^{r} h_i(\hat{x}(t))[\| GL_{i,m} \| \| y(t) \| + \| GL_{i,m} C_m \| \| \hat{x}(t) \|],$$

And

$$\delta \geq \frac{\| G\hat{x}(0) \|}{T}.$$

Proof Choose the following Lyapunov function:

$$V(t) = \frac{1}{2} s^T(t) s(t). \tag{4.12}$$

Then, the infinitesimal generator of Lyapunov function $V(t)$ along the trajectory of fuzzy observer system is given as:

$$\mathcal{L}V(t) = s^T(t)\dot{s}(t)$$

$$= s^T(t) \sum_{i=1}^{r} h_i(\hat{x}(t)) G \left[L_{i,m} C_m \theta(t) - B_m K_{i,m} \hat{x}(t) + GB_m u(t) \right].$$

$$\leq \| s(t) \| \sum_{i} h_i(\hat{x}(t))[\| GL_{i,m} \| \| y(t) \| + \| GL_{i,m} C_m \| \| \hat{x}(t) \|]$$

$$- s^T(t) \sum_{i=1}^{r} h_i(\hat{x}(t)) GB_m K_{i,m} \hat{x}(t) + s^T(t) GB_m u(t)$$

$$\tag{4.13}$$

Combined with the controller (4.11), it follows that

$$\mathcal{L}V(t) \leq -\delta \|s(t)\| = -\delta\sqrt{2}V^{\frac{1}{2}}(t). \qquad (4.14)$$

From the above inequality, we can deduce there is an instant $T^* \leq \sqrt{2V(0)}/\delta$, which satisfies that $V(t) = 0$ (i.e., $s(t) = 0$) for all $t \geq T^*$. On the other hand,

$$V(0) = \frac{1}{2}\|s(0)\|^2 = \frac{1}{2}\|G\hat{x}(0)\|^2. \qquad (4.15)$$

So, it yields $T^* \leq \dfrac{\|G\hat{x}(0)\|}{\delta}$.

Combining the δ designed in controller (4.11), we can get $T^* \leq T$. Therefore, the sliding surface $s(t) = 0$ can be reached before time T as we desired. This completes the proof.

Remark 4.4

As we can see from the above proof, the parameter δ in controller (4.11) matters whether the sliding surface can be reached before time T or not. And seeing from the proof, it is known that the bigger the value of δ is, then faster the sliding surface will be reached.

4.3.2 FINITE-TIME BOUNDEDNESS ANALYSIS DURING [0, T*]

In this section, we will demonstrate that the fuzzy observer system (4.6) by the controller (4.11) combined with the error dynamics (4.7) during the phase $[0, T^*]$ is FTB. The closed-loop system of the fuzzy observer (4.6) during the reaching phase is obtained:

$$\dot{\hat{x}}(t) = \sum_{i=1}^{r} h_i(\hat{x}(t))\left[(A_{i,m} + B_m K_{i,m})\hat{x}(t) + L_{i,m}C_m\theta(t) - B_m(GB_m)^{-1}\hat{\rho}(t)\right], \qquad (4.16)$$

where $\hat{\rho}(t) = (\rho(t) + \delta)sgn(s(t))$.

In the following, conditions are provided to guarantee the overall closed-loop system (4.16) combined with the error dynamics (4.7) is FTB.

Theorem 4.2

For a given scalar $T^* > 0$, the fuzzy system (4.16) combined with the error dynamics (4.7) is FTB with respect to $\{c_1, c_2, T^*, R, d\}$ via the controller (4.11), if there exist matrices $P_m > 0, Q_1 > 0, Q_2 > 0$, positive scalars $\beta > 0, \gamma > 0$ and $c' > 0$ such that the following conditions hold for all $m \in S$

$$Q_1 - 12a^2\gamma I \geq 0, \qquad (4.17)$$

$$Q_2 - 3a^2\gamma I \geq 0, \qquad (4.18)$$

$$\begin{bmatrix} \Theta_{i,m}^1 & H_{i,m}C_m & 0 & -P_mB_m(GB_m)^{-1} \\ * & \Theta_{i,m}^2 & P_m & 0 \\ * & * & -\beta I & 0 \\ * & * & * & -\gamma I \end{bmatrix} < 0, \qquad (4.19)$$

$$\frac{c_1\lambda_1 + \beta d(1 - e^{-\alpha T^*})/\alpha + 3\delta^2\gamma T^*}{\lambda_2} < e^{-\alpha T^*}c', \qquad (4.20)$$

where $a = \max_{i,m \in S} \| GL_{i,m}C_m \|$, $\lambda_1 = \lambda_{\max}(\bar{P}_m)$, $\lambda_2 = \lambda_{\min}(\bar{P}_m)$, $\bar{P}_m = R^{-\frac{1}{2}}P_mR^{-\frac{1}{2}}$, and

$$\Theta_{i,m}^1 = \mathrm{He}\{P_m(A_{i,m} + B_mK_{i,m})\} - \alpha P_m + Q_1 + \sum_{n=1}^{s}\pi_{mn}(h)P_n,$$

$$\Theta_{i,m}^2 = \mathrm{He}\{P_mA_{i,m} - H_{i,m}C_m\} - \alpha P_m + Q_2 + \sum_{n=1}^{s}\pi_{mn}(h)P_n,$$

in which $K_{i,m}$ is selected such that $A_{i,m} + B_mK_{i,m}$ is Hurwitz. Moreover, the observer gain matrix is given by $L_{i,m} = P_m^{-1}H_{i,m}$.

Proof: Select the following Lyapunov function:

$$V(\hat{x}(t),\theta(t),r_t) = \hat{x}^T(t)P(r_t)\hat{x}(t) + \theta^T(t)P(r_t)\theta(t)$$

$$+ \int_0^t e^{\alpha(t-s)}\hat{x}^T(s)Q_1\hat{x}(s)ds + \int_0^t e^{\alpha(t-s)}\theta^T(s)Q_2\theta(s)ds. \qquad (4.21)$$

Then, we have

$$\mathcal{L}V(\hat{x}(t),\theta(t),t) = 2\hat{x}^T(t)P_m\sum_{i=1}^{r}h_i(\hat{x}(t))\big[(A_{i,m} + B_mK_{i,m})\hat{x}(t) + L_{i,m}C_m\theta(t)$$

$$-B_m(GB_m)^{-1}\hat{\rho}(t)\big] + \hat{x}^T(t)\sum_{n=1}^{s}\pi_{mn}(h)P_n\hat{x}(t)$$

$$+ 2e^T(t)P_m\sum_{i=1}^{r}h_i(\hat{x}(t))[(A_{i,m} - L_{i,m}C_m)\theta(t) + w(t)]$$

$$+ \theta^T(t)\sum_{n=1}^{s}\pi_{mn}(h)P_n\theta(t) + \hat{x}^T(t)Q_1\hat{x}(t) + \theta^T(t)Q_2\theta(t)$$

$$+ \alpha V(t) - \alpha\hat{x}^T(t)P_m\hat{x}(t) - \alpha\theta^T(t)P_m\theta(t). \qquad (4.22)$$

Defining $P_m L_{i,m} = H_{i,m}$, based on (4.19), it holds

$$\mathcal{L}V(t) - \alpha V(t) - \beta w^T(t)w(t) - \gamma \hat{\rho}^T(t)(t)\hat{\rho}(t) \le \eta^T(t)\Theta_{i,m}\eta(t) < 0, \qquad (4.23)$$

where $\eta^T(t) = [\hat{x}^T(t) \ \theta^T(t) \ w^T(t) \ \hat{\rho}^T(t)]$. Then, it follows from (4.23) that

$$e^{-\alpha t}\mathbf{E}[\mathcal{L}V(t) - \alpha V(t)] \le e^{-\alpha t}\mathbf{E}[\beta w^T(t)w(t) + \gamma \hat{\rho}^T(t)\hat{\rho}(t)]. \qquad (4.24)$$

Integrating inequality (4.24) from 0 to t yields

$$\mathbf{E}V(t) \le e^{\alpha t}V(0) + e^{\alpha t}\mathbf{E}\int_0^t e^{-\alpha s}[\beta w^T(s)w(s) + \gamma \hat{\rho}^T(s)\hat{\rho}(s)]ds. \qquad (4.25)$$

Taking (4.21) into consideration, it holds

$$\mathbf{E}[\hat{x}^T(t)P_m\hat{x}(t)] \le e^{\alpha t}V(0) + \mathbf{E}\int_0^t e^{\alpha(t-s)}[\beta w^T(s)w(s)$$

$$+3\gamma\delta^2]ds - \mathbf{E}\int_0^t e^{\alpha(t-s)}\hat{x}^T(s)(Q_1 - 12a^2\gamma I)\hat{x}(s)]ds$$

$$-\mathbf{E}\int_0^t e^{\alpha(t-s)}e^T(s)(Q_2 - 3a^2\gamma I)e(s)]ds$$

$$\le e^{\alpha t}V(0) + \mathbf{E}\int_0^t e^{\alpha(t-s)}[\beta w^T(s)w(s) + 3\gamma\delta^2]ds, \qquad (4.26)$$

in which the fact $\hat{\rho}^T(t)\hat{\rho}(t) \le 3\delta^2 + 3a^2 \|\theta(t)\|^2 + 3(2a)^2 \|\hat{x}(t)\|^2$ is used.

Now, by denoting $\bar{P}_m = R^{-\frac{1}{2}}P_m R^{-\frac{1}{2}}$, we have

$$V(0) = \hat{x}^T(0)P_m\hat{x}(0) + e^T(0)P_m e(0)$$

$$\le \lambda_{\max}\{\bar{P}_m\}(\hat{x}^T(0)P_m\hat{x}(0) + \theta^T(0)P_m\theta(0))$$

$$\le c_1\lambda_{\max}\{\bar{P}_m\}. \qquad (4.27)$$

On the other hand, it holds that

$$\mathbf{E}[\hat{x}^T(t)P_m\hat{x}(t)] \ge \lambda_{\min}\{\bar{P}_m\}\mathbf{E}[\hat{x}^T(t)R\hat{x}(t)]. \qquad (4.28)$$

Therefore, combining (4.26)–(4.28) yields

$$\mathbf{E}[\hat{x}^T(t)R\hat{x}(t)] \le \frac{e^{\alpha T^*}[c_1\lambda_1 + \beta d(1 - e^{-\alpha T^*})/\alpha + 3\gamma\delta^2 T^*]}{\lambda_2}. \qquad (4.29)$$

By the condition in (4.20), we can conclude $\mathbf{E}[\hat{x}^T(t)R\hat{x}(t)] \le c'$ for $\forall t \in [0, T^*]$. This completes the proof.

Remark 4.5

Theorem 4.2 shows the finite-time boundedness of the fuzzy observer system (4.6) before reaching onto the sliding surface, while by Definition 4.1, it is required that $E[\hat{x}^T(t)R\hat{x}(t)] \leq c_2$ for $\forall t \in [0,T]$, which means the state of the observer system (4.5) during this phase still needs to satisfy this condition. So it is obvious that $c' \leq c_2$ should be guaranteed in Theorem 4.2.

4.3.3 FINITE-TIME BOUNDEDNESS ANALYSIS OF SLIDING MODE DYNAMICS

In the former section, we have confirmed that the observer system during the reaching phase is FTB. Thus, in the section, we will provide conditions to ensure that the sliding mode dynamics is also FTB.

Based on the sliding surface function (4.10) and the fuzzy observer system (4.6), it holds

$$\dot{s}(t) = \sum_{i=1}^{r} h_i(\hat{x}(t))G\left[L_{i,m}C_m\theta(t) - B_m K_{i,m}\hat{x}(t)\right] + GB_m u(t). \qquad (4.30)$$

According to VSC theory, when the sliding surface is reached, that is, $s(t) = 0$, $\dot{s}(t) = 0$. By $\dot{s}(t) = 0$, we have the equivalent control variable as

$$u_{eq}(t) = \sum_{i=1}^{r} h_i(\hat{x}(t))[K_{i,m}\hat{x}(t) - (GB_m)^{-1}GL_{i,m}C_m\theta(t)]. \qquad (4.31)$$

Then, substituting (4.31) into (4.6) leads to the sliding mode dynamics as follows:

$$\dot{\hat{x}}(t) = \sum_{i=1}^{r} h_i(\hat{x}(t))\left[(A_{i,m} + B_m)K_{i,m}\hat{x}(t) + I_m L_{i,m}C_m\theta(t)\right], \qquad (4.32)$$

where $I_m = I - B_m(GB_m)^{-1}G$.

Theorem 4.3

Given a scalar T^{**} satisfying $T^{**} \leq T - T^*$, the sliding mode dynamics (4.32) combined with the error dynamics is FTB with respect to $\{c_1, c_2, T^{**}, R, d\}$, if there are matrices $P_m > 0$, $H_{i,m} > 0$, scalars $\beta > 0$ and $\gamma > 0$ such that the following conditions hold for all $m \in \mathcal{S}$

$$\begin{bmatrix} \Xi_{i,m}^1 & 0 & 0 & P_m \mathcal{I}_m & 0 \\ * & \Xi_{i,m}^2 & P_m & 0 & C_m^T H_{i,m}^T \\ * & * & -\beta I & 0 & 0 \\ * & * & * & -P_m & 0 \\ * & * & * & * & -P_m \end{bmatrix} < 0, \qquad (4.33)$$

$$\frac{c'\lambda_1 + \beta d}{\lambda_2} < e^{-\alpha T^{**}} c_2, \tag{4.34}$$

where λ_1, λ_2 and c' are defined as in Theorem 4.2, and

$$\Xi_{i,m}^1 = \text{He}\{P_m(A_{i,m} + B_m K_{i,m})\} - \alpha P_m + \sum_{n=1}^{s} \pi_{mn}(h)P_n,$$

$$\Xi_{i,m}^2 = \text{He}\{P_m A_{i,m} - H_{i,m} C_m\} - \alpha P_m + \sum_{n=1}^{s} \pi_{mn}(h)P_n,$$

in which $K_{i,m}$ is selected such that $A_{i,m} + B_m K_{i,m}$ is Hurwitz. Moreover, the observer gain matrix is given by $L_{i,m} = P_m^{-1} H_{i,m}$.

Proof Consider the Lyapunov function of the form:

$$V(\hat{x}(t), \theta(t), t) = \hat{x}^T(t)P(r_t)\hat{x}(t) + \theta^T(t)P(r_t)\theta(t). \tag{4.35}$$

So it follows

$$\mathcal{L}V(\hat{x}(t), \theta(t), t) = 2\hat{x}^T(t)P_m \sum_{i=1}^{r} h_i(\hat{x}(t))\left[(A_{i,m} + B_m K_{i,m})\hat{x}(t)\right.$$

$$\left. + \mathcal{I}_m L_{i,m} C_m \theta(t)\right] + \hat{x}^T(t)\sum_{n=1}^{s} \pi_{mn}(h)P_n\hat{x}(t)$$

$$+ 2\theta^T(t)P_m \sum_{i=1}^{r} h_i(\hat{x}(t))[(A_{i,m} - L_{i,m}C_m)\theta(t)$$

$$+ w(t)] + \theta^T(t)\sum_{n=1}^{s} \pi_{mn}(h)P_n\theta(t). \tag{4.36}$$

In (4.36), we have

$$2\hat{x}^T(t)P_m \mathcal{I}_m L_{i,m} C_m \theta(t) \leq \hat{x}^T(t)P_m \mathcal{I}_m P_m^{-1} \mathcal{I}_m^T P_m \hat{x}(t)$$

$$+ \theta^T(t)C_m^T L_{i,m}^T P_m L_{i,m} C_m \theta(t). \tag{4.37}$$

Then, based on condition (4.33), it holds

$$\mathcal{L}V(t) - \alpha V(t) - \beta w^T(t)w(t) \leq \eta^T(t)\Xi_{i,m}\eta(t) < 0, \tag{4.38}$$

where $\eta^T(t) = [\hat{x}^T(t) \quad \theta^T(t) \quad w^T(t)]$.

Similar to (4.24)–(4.28), we have

$$\mathbf{E}[\hat{x}^T(t)R\hat{x}(t)] \le \frac{e^{\alpha T^{**}}[c'\lambda_1 + \beta d]}{\lambda_2}. \qquad (4.39)$$

Therefore, by condition (4.34), we can conclude $\mathbf{E}[\hat{x}^T(t)R\hat{x}(t)] \le c_2$ for $t \in [0, T^{**}]$. This completes the proof.

From Theorem 4.2 and Theorem 4.3, finite-time boundedness analysis is undertaken independently during the two phases, which is inconvenient to solve the sliding mode controller design problems simultaneously. Additionally, the existence of time-varying TRs is also a tough problem to be tackled. Thus, in the following, the overall analysis of FTB during the whole phase $[0, T]$ is given although the TRs are generally uncertain.

4.3.4 FINITE-TIME BOUNDEDNESS ANALYSIS OVER [0,T]

Theorem 4.4

Given scalar $T > 0$, the fuzzy observer system (4.5) combined with the error dynamics (4.6) is FTB with respect to $\{c_1, c_2, T, R, d\}$ by the controller (4.11), if there are matrices $P_m > 0$, $T_{mn} > 0$, $Z_{mn} > 0$, $Q_1 > 0$, $Q_2 > 0$ and $H_{i,m}$ with appropriate dimensions, scalars $\beta > 0$, $\gamma > 0$ and λ_i $(i = 1, 2)$ such that (4.17), (4.18), (4.20), (4.34) and the following conditions hold for all $m \in S$

If $m \in I_{m,k}$, $\forall l \in I_{m,uk}$, $I_{m,k} \triangleq \{k_{m,1}, k_{m,2}, \ldots, k_{m,o_1}\}$,

$$\begin{bmatrix} \mathcal{A}_{i,m}^1 & \mathcal{A}_{i,m}^2 & \mathcal{A}_{i,m}^3 \\ * & \mathcal{A}_{i,m}^4 & 0 \\ * & * & \mathcal{A}_{i,m}^5 \end{bmatrix} < 0, \qquad (4.40)$$

with

$$\mathcal{A}_{i,m}^1 = \begin{bmatrix} \mathcal{A}_{i,m}^{11} & H_{i,m}C_m & 0 \\ * & \mathcal{A}_{i,m}^{12} & 0 \\ * & * & \mathcal{A}_{i,m}^{13} \end{bmatrix},$$

$$\mathcal{A}_{i,m}^2 = \begin{bmatrix} 0 & -P_m B_m (GB_m)^{-1} & P_m \mathcal{I}_m & 0 \\ P_m & 0 & 0 & 0 \\ P_m & 0 & 0 & C_m^T H_{i,m}^T \end{bmatrix},$$

$$\mathcal{A}_{i,m}^3 = \begin{bmatrix} \mathcal{A}_{i,m}^{31} & 0 & 0 \\ * & \mathcal{A}_{i,m}^{32} & 0 \\ * & * & \mathcal{A}_{i,m}^{33} \end{bmatrix},$$

$$\mathcal{A}_{i,m}^5 = diag\{\mathcal{A}_{i,m}^{51}, \mathcal{A}_{i,m}^{52}, \mathcal{A}_{i,m}^{53}\},$$

If $m \in I_{m,uk}$, $\forall l \in I_{m,uk}$, $I_{m,k} \triangleq \{k_{m,1}, k_{m,2}, \ldots, k_{m,o2}\}$, $l \neq m$,

$$P_m - P_l \geq 0, \tag{4.41}$$

$$\begin{bmatrix} \mathcal{B}_{i,m}^1 & \mathcal{B}_{i,m}^2 & \mathcal{B}_{i,m}^3 \\ * & \mathcal{B}_{i,m}^4 & 0 \\ * & * & \mathcal{B}_{i,m}^5 \end{bmatrix} < 0, \tag{4.42}$$

with

$$\mathcal{B}_{i,m}^1 = \begin{bmatrix} \mathcal{B}_{i,m}^{11} & H_{i,m}C_m & 0 \\ * & \mathcal{B}_{i,m}^{12} & 0 \\ * & * & \mathcal{B}_{i,m}^{13} \end{bmatrix},$$

$$\mathcal{B}_{i,m}^3 = \begin{bmatrix} \mathcal{B}_{i,m}^{31} & 0 & 0 \\ * & \mathcal{B}_{i,m}^{32} & 0 \\ * & * & \mathcal{B}_{i,m}^{33} \end{bmatrix},$$

and

$$\mathcal{B}_{i,m}^5 = diag\left\{\mathcal{B}_{i,m}^{51}, \mathcal{B}_{i,m}^{52}, \mathcal{B}_{i,m}^{53}\right\},$$

$$\lambda_1 R \leq P_m \leq \lambda_2 R, \tag{4.43}$$

where

$$\mathcal{A}_{i,m}^{11} = He\{P_m(A_{i,m} + B_m K_{i,m})\} - \alpha P_m + Q_1 + \sum_{n \in I_{m,k}} \left[\frac{(\lambda_{mn})^2}{4} T_{mn} + \pi_{mn}(P_n - P_l)\right],$$

$$\mathcal{A}_{i,m}^{12} = He\{P_m A_{i,m} - H_{i,m}C_m\} - \alpha P_m + Q_2 + \sum_{n \in I_{m,k}} \left[\frac{(\lambda_{mn})^2}{4} T_{mn} + \pi_{mn}(P_n - P_l)\right],$$

$$\mathcal{A}_{i,m}^{13} = He\{P_m A_{i,m} - H_{i,m}C_m\} - \alpha P_m + \sum_{n \in I_{m,k}} \left[\frac{(\lambda_{mn})^2}{4} T_{mn} + \pi_{mn}(P_n - P_l)\right],$$

$$\mathcal{A}_m^{31} = \mathcal{A}_m^{32} = \mathcal{A}_m^{33} = [(P_{k_{m,1}} - P_l) \ \ldots \ (P_{k_{m,o1}} - P_l)],$$

$$\mathcal{A}_{i,m}^4 = diag\{-\beta I, -\gamma I, -P_m, -P_m\},$$

$$\mathcal{A}_{i,m}^{51} = \mathcal{A}_{i,m}^{52} = \mathcal{A}_{i,m}^{53} = diag\{-T_{mk_{m,1}}, \ldots, -T_{mk_{m,o1}}\},$$

$$\mathcal{B}_{i,m}^{11} = \mathrm{He}\{P_m(A_{i,m} + B_m K_{i,m})\} - \alpha P_m + Q_1 + \sum_{n \in I_{m,k}} \left[\frac{(\lambda_{mn})^2}{4} Z_{mn} + \pi_{mn}(P_n - P_l) \right],$$

$$\mathcal{B}_{i,m}^{12} = \mathrm{He}\{P_m A_{i,m} - H_{i,m} C_m\} - \alpha P_m + Q_2 + \sum_{n \in I_{m,k}} \left[\frac{(\lambda_{mn})^2}{4} Z_{mn} + \pi_{mn}(P_n - P_l) \right],$$

$$\mathcal{B}_{i,m}^{13} = \mathrm{He}\{P_m A_{i,m} - H_{i,m} C_m\} - \alpha P_m + \sum_{n \in I_{m,k}} \left[\frac{(\lambda_{mn})^2}{4} Z_{mn} + \pi_{mn}(P_n - P_l) \right],$$

$$\mathcal{B}_m^{31} = \mathcal{A}_m^{32} = \mathcal{A}_m^{33} = [(P_{k_{m,1}} - P_l) \ \cdots \ (P_{k_{m,o2}} - P_l)],$$

$$\mathcal{B}_{i,m}^2 = \mathcal{A}_{i,m}^2, \mathcal{B}_{i,m}^4 = \mathcal{A}_{i,m}^4,$$

$$\mathcal{B}_{i,m}^{51} = \mathcal{B}_{i,m}^{52} = \mathcal{B}_{i,m}^{53} = diag\{-Z_{mk_{m,1}}, \ldots, -Z_{mk_{m,o2}}\},$$

in which $K_{i,m}$ is selected such that $A_{i,m} + B_m K_{i,m}$ is Hurwitz. Moreover, the state observer gain matrices are given by $L_{i,m} = P_m^{-1} H_{i,m}$.

Proof: Define

$$\bar{\Gamma}_{i,m} = \Gamma_{i,m} + diag\{\Lambda_m, \Lambda_m, \Lambda_m, 0, 0, 0, 0\}, \tag{4.44}$$

where $\Lambda_m = \displaystyle\sum_{n=1}^s \pi_{mn}(h) P_n$, and

$$\Gamma_{i,m} = \begin{bmatrix} \Gamma_{i,m}^1 & H_{i,m} C_m & 0 & 0 & \Gamma_{i,m}^4 & P_m \mathcal{I}_m & 0 \\ * & \Gamma_{i,m}^2 & 0 & P_m & 0 & 0 & 0 \\ * & * & \Gamma_{i,m}^3 & P_m & 0 & 0 & C_m^T H_{i,m}^T \\ * & * & * & -\beta I & 0 & 0 & 0 \\ * & * & * & * & -\gamma I & 0 & 0 \\ * & * & * & * & * & -P_m & 0 \\ * & * & * & * & * & * & -P_m \end{bmatrix}$$

with

$$\Gamma_{i,m}^1 = \mathrm{He}\{P_m(A_{i,m} + B_m K_{i,m})\} - \alpha P_m + Q,$$

$$\Gamma_{i,m}^2 = \mathrm{He}\{P_m A_{i,m} - H_{i,m} C_m\} - \alpha P_m + Q,$$

$$\Gamma_{i,m}^3 = \mathrm{He}\{P_m A_{i,m} - H_{i,m} C_m\} - \alpha P_m,$$

$$\Gamma_{i,m}^4 = -P_m B_m (GB_m)^{-1}.$$

As we can see, $\overline{\Gamma}_{i,m} < 0$ guarantees (4.19) and (4.33) hold.

Next, we consider the term Λ_m in $\overline{\Gamma}_{i,m}$. Let us consider the following two cases.

Case I: $m \in I_{m,k}$.

First, denote $\lambda_{m,k} \triangleq \sum_{n \in I_{m,k}} \pi_{mn}(h)$. Since $I_{m,uk} \neq \varnothing$, it holds that $\lambda_{m,k} < 0$. Notice that $\sum_{n=1}^{s} \pi_{mn}(h) P_n$ can be represented as

$$\Lambda_m = \left(\sum_{n \in I_{m,k}} + \sum_{n \in I_{m,uk}} \right) \pi_{mn}(h) P_n$$

$$= \sum_{n \in I_{m,k}} \pi_{mn}(h) P_n - \lambda_{m,k} \sum_{n \in I_{m,uk}} \frac{\pi_{mn}(h)}{-\lambda_{m,k}} P_n. \qquad (4.45)$$

It is obvious that $0 \leq \pi_{mn}(h)/-\lambda_{m,k} \leq 1 \ (n \in I_{m,uk})$ and $\sum_{n \in I_{m,uk}} \frac{\pi_{mn}(h)}{-\lambda_{m,k}} = 1$. So for $\forall l \in I_{m,uk}$, there is

$$\overline{\Gamma}_{i,m} = \sum_{n \in I_{m,uk}} \frac{\pi_{mn}(h)}{-\lambda_{m,k}} \left[\Gamma_{i,m} + diag\{ \sum_{n \in I_{m,k}} \pi_{mn}(h)(P_n - P_l), \right.$$

$$\left. \sum_{n \in I_{m,k}} \pi_{mn}(h)(P_n - P_l), \sum_{n \in I_{m,k}} \pi_{mn}(h)(P_n - P_l), 0, 0, 0, 0 \} \right]. \qquad (4.46)$$

Therefore, for $0 \leq \pi_{mn}(h) \leq -\lambda_{m,k}$, $\overline{\Gamma}_{i,m} < 0$ is equivalent to

$$\Gamma_{i,m} + diag\{ \sum_{n \in I_{m,k}} \pi_{mn}(h)(P_n - P_l), \sum_{n \in I_{m,k}} \pi_{mn}(h)(P_n - P_l),$$

$$\sum_{n \in I_{m,k}} \pi_{mn}(h)(P_n - P_l), 0, 0, 0, 0, \} < 0. \qquad (4.47)$$

In formula (4.47), it is true that

$$\sum_{n \in I_{m,k}} \pi_{mn}(h)(P_n - P_l) = \sum_{n \in I_{m,k}} \pi_{mn}(P_n - P_l) + \sum_{n \in I_{m,k}} \Delta\pi_{mn}(h)(P_n - P_l). \qquad (4.48)$$

Then, by virtue of Lemma 1.4 and for any $T_{mn} > 0$, it follows that

$$\sum_{n \in I_{m,k}} \Delta\pi_{mn}(h)(P_n - P_l) = \sum_{n \in I_{m,k}} \left[\frac{1}{2} \Delta\pi_{mn}(h)((P_n - P_l) + (P_n - P_l)) \right]$$

$$\leq \sum_{n \in I_{m,k}} \left[\frac{(\lambda_{mn})^2}{4} T_{mn} + (P_n - P_l)(T_{mn})^{-1} (P_n - P_l)^T \right]. \qquad (4.49)$$

Combining (4.45)–(4.49), then by applying Schur complement, we can see that (4.40) guarantees $\overline{\Gamma}_{i,m} < 0$ when $m \in I_{m,k}$.

CaseII : $m \in I_{m,uk}$.

Similarly, denote $\lambda_{m,k} \triangleq \sum_{n \in I_{m,k}} \pi_{mn}(h)$. Since $I_{m,k} \neq \varnothing$, it holds that $\lambda_{m,k} > 0$. Now, $\sum_{n=1}^{s} \pi_{mn}(h)P_n$ can be represented as

$$
\begin{aligned}
\Lambda_m &= \sum_{n \in I_{m,k}} \pi_{mn}(h)P_n + \pi_{mm}(h)P_m + \sum_{n \in I_{m,uk},n \neq m} \pi_{mn}(h)P_n \\
&= \sum_{n \in I_{m,k}} \pi_{mn}(h)P_n + \pi_{mm}(h)P_m - (\pi_{mm}(h) + \lambda_{m,k}) \sum_{n \in I_{m,uk},n \neq m} \frac{\pi_{mn}(h)P_n}{-\pi_{mm}(h) - \lambda_{m,k}}
\end{aligned}
\tag{4.50}
$$

and it is obvious that $0 \leq \pi_{mn}(h)/-\pi_{mm}(h) - \lambda_{m,k} \leq 1$ $(n \in I_{m,uk})$ and $\sum_{n \in I_{m,uk},n \neq m} \frac{\pi_{mn}(h)}{-\pi_{mn}(h) - \lambda_{m,k}} = 1$. So for $\forall l \in I_{m,uk}, l \neq m$,

$$
\overline{\Gamma}_{i,m} = \sum_{n \in I_{m,uk},n \neq m} \frac{\pi_{mn}(h)}{-\pi_{mm}(h) - \lambda_{m,k}} \left[\Gamma_{i,m} + diag\{Y_m, Y_m, Y_m, 0,0,0,0\} \right], \tag{4.51}
$$

where $\Upsilon_m = \pi_{mm}(h)(P_m - P_l) + \sum_{n \in I_{m,k}} \pi_{mn}(h)(P_n - P_l)$.

Therefore, for $0 \leq \pi_{mn}(h) \leq -\pi_{mm}(h) - \lambda_{m,k}, \overline{\Gamma}_{i,m} < 0$ is equivalent to

$$
\Gamma_{i,m} + diag\{\Upsilon_m, \Upsilon_m, \Upsilon_m, 0,0,0,0\} < 0. \tag{4.52}
$$

Since $\pi_{mm}(h) < 0$, (4.52) holds if we have

$$
\left\{
\begin{aligned}
&P_m - P_l \geq 0, \\
&\Gamma_{i,m} + diag\{ \sum_{n \in I_{m,k}} \pi_{mn}(h)(P_n - P_l), \sum_{n \in I_{m,k}} \pi_{mn}(h)(P_n - P_l), \\
&\sum_{n \in I_{m,k}} \pi_{mn}(h)(P_n - P_l),0,0,0,0\} < 0.
\end{aligned}
\right.
\tag{4.53}
$$

Also, as in (4.48) and (4.49), for any $Z_{mn} > 0$, we have

$$
\begin{aligned}
\sum_{n \in I_{m,k}} \pi_{mn}(h)(P_n - P_l) &\leq \sum_{n \in I_{m,k}} \pi_{mn}(P_n - P_l) \\
&+ \sum_{n \in I_{m,k}} \left[\frac{(\lambda_{mn})^2}{4} Z_{mn} + (P_n - P_l)(Z_{mn})^{-1}(P_n - P_l)^T \right].
\end{aligned}
\tag{4.54}
$$

Combining (4.50)–(4.54), we know that (4.41) and (4.42) guarantee $\overline{\Gamma}_{i,m} < 0$ by applying Schur complement when $m \in I_{m,uk}$. In summary, the fuzzy observer system (4.5) is FTB over [0,T] despite the existence of generally uncertain TRs from the above analysis. This completes the proof.

Remark 4.6

Sufficient conditions in Theorem 4.4 have been developed such that the overall fuzzy observer system (4.5) is FTB over the whole phase [0,T], which guarantees Theorem 4.2 and Theorem 4.3 hold simultaneously. So, the overall fuzzy observer system is FTB during both phases. Also, a set of newly developed linear matrix inequality (LMI) conditions in Theorem 4 have been provided to ensure FTB of the overall closed-loop system and error dynamic system with generally uncertain TRs. The provided method in dealing with time-varying TRs has not shown up yet, which is one of the novelties of this paper. Also, we can see that the parameters a and L_i are interconnected in conditions (4.17)–(4.20). Thus, when solving these LMIs, the "For-Cycle" on β or γ step by step to obtain feasible solutions is used. In the following example, we will choose γ for cycling. The algorithm is adopted as follows:

Step 1: set $\gamma = 0.0001$ and a step width $d = 0.0001$;

Step 2: check (4.39)–(4.41) to see if there are feasible solutions; if not, increase γ by d, then repeat this step;

Step 3: If (4.39)–(4.41) have feasible solutions, then we can calculate a. Substitute a into (4.17) and (4.18) to check if these two conditions hold; if not, return to step 2. Otherwise, exit.

4.4 NUMERICAL EXAMPLE

In the following, a numerical example is provided to demonstrate the validity of the finite-time fuzzy SMC strategy for the model (4.1). Consider a single-link robot arm in Ref. [9] with the dynamic equation described by

$$\ddot{\vartheta}(t) = -\frac{MgL}{J}\sin(\vartheta(t)) - \frac{D(t)}{J}\dot{\vartheta}(t) + \frac{1}{J}u(t),$$

where $\vartheta(t)$ and $u(t)$ are the model angle position of the arm and control input, respectively; M, J, g, L and $D(t)$ are the model mass of the payload, the moment of inertia, the acceleration of gravity, the length of the arm and the coefficient of viscous friction, respectively. Among these parameters, $g = 9.81$, $L = 0.5$ and $D(t) = D_0 = 2$, which are time-invariant. Moreover, the three different modes of parameters M and J are shown in Table 4.1.

TABLE 4.1

Modes and Values of the Parameters M and J

Mode m	Parameter M	Parameter J
1	1	1
2	1.5	2
3	2	2.5

The generally uncertain TR matrix that regulates the operation modes is given as follows:

$$\begin{bmatrix} -0.5 + \Delta\pi_{11}(h) & ? & ? \\ ? & ? & 1.0 + \Delta\pi_{23}(h) \\ 1.5 + \Delta\pi_{31}(h) & ? & -2.0 + \Delta\pi_{33}(h) \end{bmatrix}.$$

Denote $x_1(t) = \vartheta(t)$ and $x_2(t) = \dot{\vartheta}(t)$. Similar to the method in Ref. [10], $\sin(x_1(t))$ is represented by

$$\sin(x_1(t)) = h_1(x_1(t))x_1(t) + \beta h_2(x_1(t))x_1(t),$$

where $\beta = 0.01 / \pi$, $h_1(x_1(t))$, $h_2(x_1(t)) \in [0,1]$ and $h_1(x_1(t)) + h_2(x_1(t)) = 1$. Thus, the membership functions $h_1(x_1(t))$ and $h_2(x_1(t))$ are given correspondingly as:

$$h_1(x_1(t)) = \begin{cases} \dfrac{\sin(x_1(t)) - \beta x_1(t)}{x_1(t)(1-\beta)} & x_1(t) \neq 0 \\ 1, & x_1(t) = 0 \end{cases}$$

$$h_2(x_1(t)) = \begin{cases} \dfrac{x_1(t) - \sin(x_1(t))}{x_1(t)(1-\beta)} & x_1(t) \neq 0 \\ 0, & x_1(t) = 0 \end{cases}.$$

It is evident from the above membership functions that if $x_1(t) = 0$ rad, then $h_1(x_1(t)) = 1$, $h_2(x_1(t)) = 0$, and if $x_1(t) = \pi$ rad or $x_1(t) = -\pi$ rad, then $h_1(x_1(t)) = 0$, $h_2(x_1(t)) = 1$. Thus, the state-space representation of single-link robot arm is expressed by the following two-rule system.

Plant Rule 1: IF $x_1(t)$ is "about 0 rad,"
THEN

$$\begin{cases} \dot{x}(t) = A_{1,m}x(t) + B_m u(t) \\ y(t) = C_m x(t). \end{cases}$$

Plant Rule 2: **IF** $x_1(t)$ is "about π rad or $-\pi$ rad,"
    THEN

$$\begin{cases} \dot{x}(t) = A_{2,m}x(t) + B_m u(t) \\ y(t) = C_m x(t). \end{cases}$$

where $x(t) = [x_1^T(t) \ x_2^T(t)]^T$, and

$$A_{1,1} = \begin{bmatrix} 0 & 1 \\ -gL & -D_0 \end{bmatrix}, B_1 = \begin{bmatrix} 0 \\ 1 \end{bmatrix},$$

$$A_{1,2} = \begin{bmatrix} 0 & 1 \\ -0.75gL & -0.5D_0 \end{bmatrix}, B_2 = \begin{bmatrix} 0 \\ 0.5 \end{bmatrix},$$

$$A_{1,3} = \begin{bmatrix} 0 & 1 \\ -0.8gL & -0.4D_0 \end{bmatrix}, B_3 = \begin{bmatrix} 0 \\ 0.4 \end{bmatrix},$$

$$A_{2,1} = \begin{bmatrix} 0 & 1 \\ -\beta gL & -D_0 \end{bmatrix}, C_1 = [0.1 \ 0.1],$$

$$A_{2,2} = \begin{bmatrix} 0 & 1 \\ -0.75\beta gL & -0.5D_0 \end{bmatrix}, C_2 = [0.1 \ 0.2],$$

$$A_{2,3} = \begin{bmatrix} 0 & 1 \\ -0.8\beta gL & -0.4D_0 \end{bmatrix}, C_3 = [0.1 \ 0].$$

As we can see, the system (4.1) describes this model well. So we can use the conditions in Theorem 4.4 to check the effectiveness of the proposed finite-time SMC strategy. Let $G = [0 \ 0.5]$ such that GB_m is nonsingular, $K_{1,1} = [-5 \ -3]$, $K_{1,2} = [-3 \ -2]$, $K_{1,3} = [-4 \ -2]$, $K_{2,1} = [-3 \ -1]$, $K_{2,2} = [-6 \ -3]$, $K_{2,3} = [-7 \ -6]$, and $\Delta\pi_{mn}(h) \le \lambda_{mn} = |0.1 * \pi_{mn}|$. Moreover, the following parameters are selected: $c_1 = 1$, $c' = 3, c_2 = 5, T^* = 0.5$ s, $T = 1.5$ s, $d = 3, \alpha = 4.2$, and $R = I$. By checking these LMIs, we have the following feasible solutions:

$$P_1 = \begin{bmatrix} 2.0057 & -0.0205 \\ -0.0205 & 0.4083 \end{bmatrix}, P_2 = \begin{bmatrix} 2.3467 & -0.1165 \\ -0.1165 & 0.4691 \end{bmatrix},$$

$$P_3 = \begin{bmatrix} 2.2358 & -0.2323 \\ -0.2323 & 0.4703 \end{bmatrix}, T_{1,1} = \begin{bmatrix} 7.2947 & -0.0204 \\ -0.0204 & 7.2370 \end{bmatrix},$$

$$T_{3,1} = \begin{bmatrix} 7.2149 & -0.0622 \\ -0.0622 & 6.4996 \end{bmatrix}, T_{3,3} = \begin{bmatrix} 7.1301 & -0.0508 \\ -0.0508 & 5.9953 \end{bmatrix},$$

$$Z_{2,3} = \begin{bmatrix} 7.3405 & -0.0428 \\ -0.0428 & 6.9818 \end{bmatrix}, H_{1,1} = \begin{bmatrix} 7.0429 \\ 2.0460 \end{bmatrix},$$

$$Q_1 = \begin{bmatrix} 1.8990 & -0.0015 \\ -0.0015 & 1.5671 \end{bmatrix}, H_{1,2} = \begin{bmatrix} 3.1190 \\ 0.5359 \end{bmatrix},$$

$$Q_2 = \begin{bmatrix} 0.8269 & -0.2451 \\ -0.2451 & 0.6076 \end{bmatrix}, H_{1,3} = \begin{bmatrix} 11.8216 \\ 0.9701 \end{bmatrix},$$

$$H_{2,1} = \begin{bmatrix} 8.5544 \\ 0.9373 \end{bmatrix}, H_{2,2} = \begin{bmatrix} 4.1677 \\ 0.8162 \end{bmatrix}, H_{2,3} = \begin{bmatrix} 11.4473 \\ 3.9121 \end{bmatrix},$$

$$\beta = 1.5375, \gamma = 0.3844.$$

Thus, we can get the observer gains

$$L_{1,1} = \begin{bmatrix} 3.5645 \\ 5.1899 \end{bmatrix}, L_{1,2} = \begin{bmatrix} 1.4031 \\ 1.4908 \end{bmatrix}, L_{1,3} = \begin{bmatrix} 5.7993 \\ 4.9272 \end{bmatrix},$$

$$L_{2,1} = \begin{bmatrix} 4.2908 \\ 2.5108 \end{bmatrix}, L_{2,2} = \begin{bmatrix} 1.8856 \\ 2.2082 \end{bmatrix}, L_{2,3} = \begin{bmatrix} 6.3080 \\ 11.4338 \end{bmatrix}.$$

Given the initial conditions, $x(0) = [0.1\pi \quad -0.5]^T$ and $\hat{x}(0) = [0.05\pi \quad -0.25]^T$. The tuning scalar is chosen as $\delta = 0.1179$. Also, in order to reduce the chattering effect of switching signals, $\text{sgn}(s(t))$ is replaced by $\dfrac{s(t)}{\| s(t) \| + 0.01}$. The simulation results are presented in Figures 4.1–4.5. Figure 4.1 plots one possible jumping modes. Figures 4.2 and 4.3 give the state response of the observer system and error dynamics, respectively. Figure 4.4 describes the sliding surface function and control input. Figure 4.5 presents the comparison of sliding surface function with different δ.

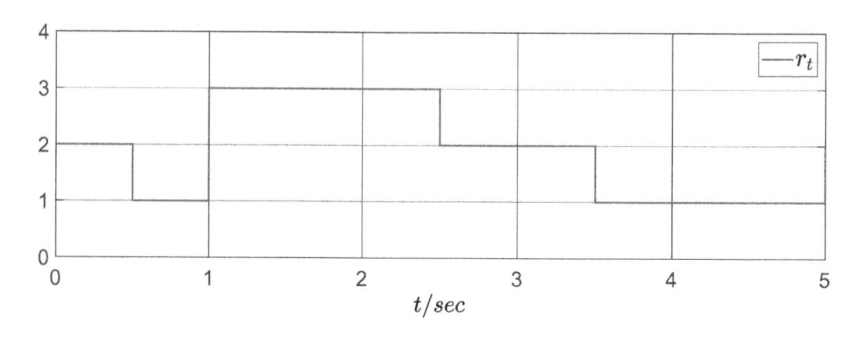

FIGURE 4.1 Jumping modes.

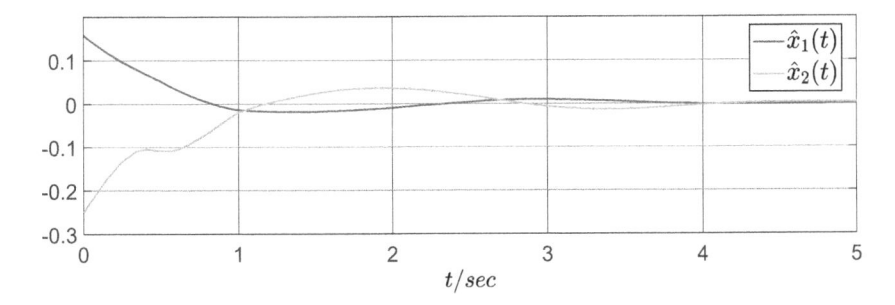

FIGURE 4.2 State response observer system $\hat{x}(t)$.

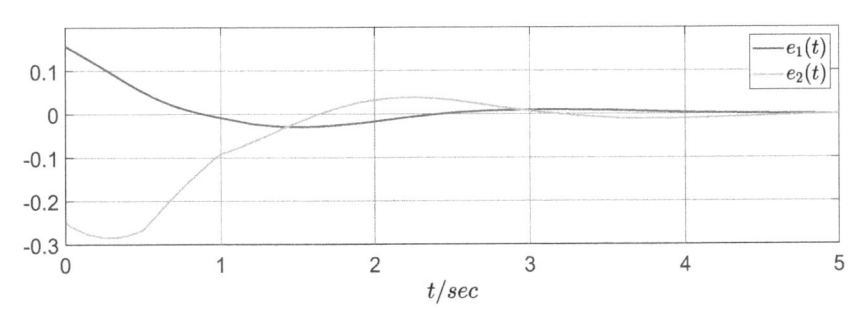

FIGURE 4.3 Estimated errors.

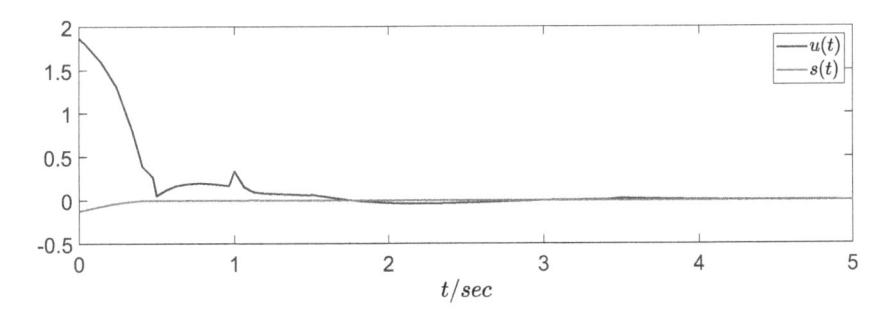

FIGURE 4.4 Response of control input $u(t)$ and sliding surface function $s(t)$.

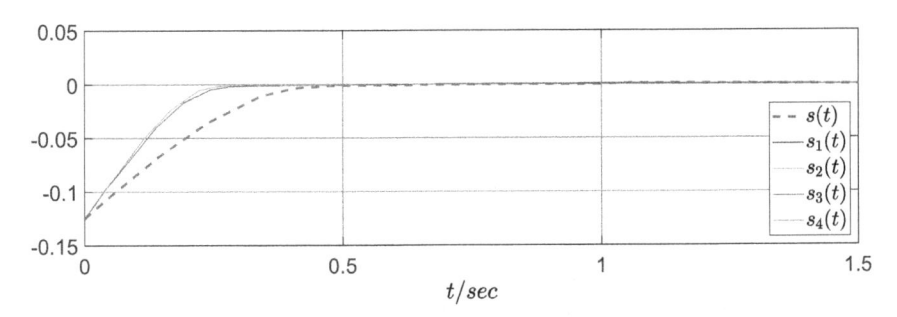

FIGURE 4.5 Response of sliding surface function $s(t)$ with different δ.

Remark 4.7

It is shown from Figure 4.4 the sliding surface $s(t) = 0$ can be reached around $T^* = 0.5$ s as we desired. And, in accordance with the theoretic analysis in Remark 4.4, the reaching response is faster as we increase the value of δ, Figure 4.5 shows the response of $s(t)$ with $\delta = 0.15$ and $s_i(t)$ $(i = 1,\ldots,4)$ with $\delta = 0.3$, respectively. But it is worth pointing out that this may not always be true because δ only limits the upper bound of the reaching time and the system may run differently under different jumping modes.

4.5 CONCLUSION

In this chapter, an observer-based finite-time SMC strategy has been proposed for S-MJSs with immeasurable premise variables via fuzzy approach. An integral sliding surface has been constructed on the observer space, and observer-based sliding mode controller has been designed to ensure finite-time reachability of the predefined sliding surface. Then, finite-time boundedness analysis off and on the sliding surface has been conducted and sufficient conditions have been established to ensure the overall closed-loop system is FTB during $[0,T]$. Lastly, a practical example has been given to demonstrate the validity of the proposed method.

REFERENCES

[1] S. He, Q. Ai, C. Ren, J. Dong and F. Liu, Finite-time resilient controller design of a class of uncertain nonlinear systems with time-delays under asynchronous switching, *IEEE Transactions on Systems, Man, and Cybernetics: Systems*, Vol. 49, No. 2, pp. 281–286, 2019.

[2] P. Dorato, Short-time stability in linear time-varying systems, *Proc. IRE Int. Convent. Rec.*, pp. 83–87, 1961.

[3] F. Amato, M. Ariola and P. Dorato, Finite-time control of linear systems subject to parametric uncertainties and disturbances, *Automatica*, Vol. 37, No. 7, pp. 1459–1463, 2001.

[4] J. Song, Y. Niu and Y. Zou, Finite-time stabilization via sliding mode control, *IEEE Transactions on Automatic Control*, Vol. 62, No. 3, pp. 1478–1483, 2017.

[5] S. He, J. Song and F. Liu, Robust finite-time bounded controller design of time-delay conic nonlinear systems using sliding mode control strategy, *IEEE Transactions on Systems, Man, and Cybernetics: Systems*. Vol. 48, No. 11, pp. 1863–1873, 2018.

[6] L. Gao, X. Jiang and D. Wang, Observer-based robust finite-time H_∞ sliding mode control for Markovian switching systems with mode- dependent time-varying delay and incompletet ransition rate, *ISA Transactions*, Vol. 61, pp. 29–48, 2016.

[7] M. Chadli, and H. R. Karimi, Robust observer design for unknown inputs Takagi-Sugeno models, *IEEE Transactions on Fuzzy Systems*, Vol. 21, No. 1, pp. 158–164, 2013.

[8] J. Yoneyama, Output feedback control for fuzzy systems with immeasurable premise variables, *IEEE International Conference on Fuzzy Systems*. IEEE, pp. 802–807, 2009.

[9] H. N. Wu and K. Y. Cai, Mode-independent robust stabilization for uncertain Markovian jump nonlinear systems via fuzzy control, *IEEE Transactions on Systems, Man, and Cybernetics: Systems*, Vol. 36, No. 3, pp. 509–519, 2005.

[10] K. Tanaka and T. Kosaki, Design of a stable fuzzy controller for an articulated vehicle, *IEEE Transactions on Systems, Man, and Cybernetics: Part B, Cybernetics*, Vol. 2, No. 3, pp. 552–558, June 1997.

5 Adaptive Fuzzy Sliding Mode Control of Semi-Markovian Jump Systems

5.1 INTRODUCTION

In the application field of engineering and science, adaptive control has been a popular control methodology attracting increasing attention due to its unique capabilities to accommodate system parametric, structural, and environmental uncertainties; adaptive control can modify its own characteristics to adapt to changes in the dynamic characteristics of objects and disturbances caused by payload variations or system aging, component failures and external disturbances. On the other hand, as mentioned before, sliding mode control (SMC) has been proven to be one of the most effective approaches for nonlinear systems. It has been widely implemented in practical systems due to the attractive properties of SMC models: for example, fast response and good transient performance, strong robustness to system perturbations and uncertainties. In general, two steps are involved in a SMC scheme: *sliding phase* synthesis and *reaching phase* synthesis. In the past few years, quite a few applications of T-S fuzzy SMC methods in literature and engineering have been proposed. What is worth mentioning is that these significant results are obtained based on integral SMC (ISMC) method, which is very popular due to its advantages that the reaching phase can be eliminated for traditional SMC method; thus, the robustness can be achieved through the whole system response. In view of this, many good research works regarding ISMC of T-S fuzzy models have been reported: in Ref. [1], the fuzzy approach was applied to stochastic time-delay model with parametric uncertainties and unknown nonlinearities. In Refs. [2,3], a novel dynamic robust H_∞ ISMC method was proposed for stochastic T-S fuzzy time-delay systems. In Ref. [4], the fuzzy ISMC approach was proposed for the stabilization of stochastic descriptor T-S fuzzy systems. And, the singularly perturbed systems with application to electric circuit were studied via fuzzy ISMC in Ref. [5]; for more references, see Refs. [6–15]. On the other hand, adaptive sliding mode controller can also provide some advantages, such as the control objective achievement despite large-scale plant uncertainties. This means the proposed adaptive controller can handle more than just constant unknown parameters: for example, time-varying parameters, unmodeled dynamics, noise and disturbance inputs, etc. Therefore, the adaptive-based ISMC for dynamical systems could be very interesting and attractive.

In practical systems, the knowledge of states is important. However, in most cases, the state variables are usually unavailable due to the fact that installing all the sensors needed may not be realistic or the costs may be too high. In such cases, the Luenberger observer can be applied to estimate the unmeasurable states and plays an important role in designing the observer-based controller. But in the observer design of T-S fuzzy systems, many papers select the premise variables of fuzzy observer, which are the same as those of T-S fuzzy model of the plant [16,17]. However, in many practical problems, the premise variables may depend on the unmeasurable state variables. So, to have a real sliding mode observer design, the premise variables of the fuzzy observer depending on the estimated state variables by the fuzzy observer must be considered. So far, we have witnessed some pioneer works done this issue, such as the robust observer design for unknown input T-S models in Ref. [18] where the case of unmeasured decision variables was studied. Based on this idea, this paper is to further investigate the adaptive ISMC of fuzzy systems in the case of unmeasurable premise variables depending on the state observer.

Based on the above analysis, we can list the following issues to be tackled: (1) how to design an observer for T-S fuzzy systems that the premise variables dependent on the system variables are not available; (2) how to propose an integral sliding surface on the estimation space that suits the system with no constraint on input matrices; (3) how to give less general conservative conditions for stochastic stability analysis of semi-Markovian jump T-S fuzzy systems in the sense of generally uncertain TRs; and (4) how to design an adaptive sliding mode controller such that the reaching condition can be satisfied and the sliding mode has good stability property. Therefore, an opening challenge that how to synthesize the problem of observer-based ISMC for T-S fuzzy semi-Markovian jump systems (S-MJSs) with immeasurable premise variables while taking all the aforementioned questions into consideration simultaneously deserves our attention. However, to the best of our knowledge, no results have been reported on this topic in the literature yet.

Therefore, this chapter will concern with the problem of observer-based fuzzy ISMC for a class of continuous-time nonlinear semi-Markovian jump T-S fuzzy systems, in which the premise variables are not measurable. A novel integral sliding surface incorporated with transformed input matrices to better accommodate the characteristic of T-S fuzzy models will be proposed on estimation space, and feasible easy-checking linear matrix inequality (LMI) conditions to ensure stochastic stability of the overall closed-loop system with generally uncertain TRs will be established.

5.2   SYSTEM DESCRIPTION

Consider the following nonlinear semi-Markovian jump T-S fuzzy system on the probability space $(\Omega, \mathcal{F}, \mathcal{P})$:

Plant Rule i : **IF** $x_1(t)$ is M_{i1} **and** $x_2(t)$ is M_{i2} **and** $\cdots$ **and** $x_n(t)$ is M_{in}
THEN

$$
\begin{cases}
\dot{x}(t) = A_i(r_t)x(t) + B_i(r_t)(u(t) + f(x(t),t)) \\
y(t) = C(r_t)x(t) \\
x(0) = \varphi(0)
\end{cases}
\tag{5.1}
$$

where $x(t) = [x_1(t) \; x_2(t) \cdots \; x_n(t)]^T \in \mathbb{R}^n$ denotes the state vector; $x_1(t), \ldots, x_n(t)$ are also selected as premise variables; $M_{ij} (i = 1,2\ldots,r; j = 1,2\ldots,n)$ are the fuzzy sets. $u(t) \in \mathbb{R}^m$ is the control input; $y(t) \in \mathbb{R}^p$ is the controlled output. $A_i(r_t)$, $B_i(r_t)$ and $C(r_t)$ are the system matrices with compatible dimensions. $f(x(t),t)$ is the local lumped uncertainty of the plant.

$\{r_t, t \geq 0\}$ is a continuous-time semi-Markovian process taking discrete values in a finite set $S = \{1,2,\ldots,s\}$ with generator given by

$$\Pr\{r_{t+h} = n \mid r_t = m\} = \begin{cases} \pi_{mn}(h)h + o(h), & m \neq n, \\ 1 + \pi_{mm}(h)h + o(h), & m = n, \end{cases} \tag{5.2}$$

where $h > 0$ and $\lim_{h \to 0} o(h)/h = 0$, $\pi_{mn}(h) > 0, m \neq n$, is the transition rate from mode m at time t to mode n at time $t + h$, and $\pi_{mm}(h) = -\sum_{n \neq m} \pi_{mn}(h) < 0$ for each $n \in S$.

In this chapter, from a practical point of view, more generally uncertain TRs are investigated and it is considered that the transition rate (TR) $\pi_{mn}(h)$ may satisfy one of the following two conditions:

    I. $\pi_{mn}(h)$ is completely unknown;
    II. $\pi_{mn}(h)$ is not exactly known but upper- and lower-bounded.

For the case (II), for instance, it is assumed $\pi_{mn}(h) \in [\underline{\pi}_{mn}, \bar{\pi}_{mn}]$, in which $\underline{\pi}_{mn}$ and $\bar{\pi}_{mn}$ are the known real constants representing the lower and upper bounds of $\pi_{mn}(h)$, respectively. In view of this, we further denote $\pi_{mn}(h) \triangleq \pi_{mn} + \Delta\pi_{mn}(h)$, in which $\pi_{mn} = \frac{1}{2}(\underline{\pi}_{mn} + \bar{\pi}_{mn})$ and $|\Delta\pi_{mn}(h)| \leq \lambda_{mn}$ with $\lambda_{mn} = \frac{1}{2}(\bar{\pi}_{mn} - \underline{\pi}_{mn})$. So the TR matrix with three jumping modes may be described as

$$\begin{bmatrix} \pi_{11} + \Delta\pi_{11}(h) & ? & \pi_{13} + \Delta\pi_{13}(h) \\ ? & ? & \pi_{23} + \Delta\pi_{23}(h) \\ ? & \pi_{32} + \Delta\pi_{32}(h) & ? \end{bmatrix}, \tag{5.3}$$

where "?" is the description of unknown TRs. For brevity, $\forall \; m \in S$, let $I_m = I_{m,k} \cup I_{m,uk}$, where

$$I_{m,k} \triangleq \{n : \pi_{mn} \text{ can be determined for } n \in S\},$$

$$I_{m,uk} \triangleq \{n : \pi_{mn} \text{ is not known for } n \in S\}.$$

More generally, here, it assumes that both $I_{m,k} \neq \varnothing$ and $I_{m,uk} \neq \varnothing$. Thus, we can denote the following set:

$$I_{m,k} \triangleq \{k_{m,1}, k_{m,2}, \ldots, k_{m,o}\} \quad 1 \leq o < s,$$

where $k_{m,s}(s \in \{1,2,\ldots,o\})$ represents the index of s-th element in the m-th row of the TR matrix.

Remark 5.1

It is noticed that some other interesting issues were also proposed with respect to the semi-Markov models, where the sojourn time of each mode was set lower-bounded in Ref. [19]. When the modes information cannot be obtained synchronously, a detector that provides estimated values of the system modes was provided in Ref. [20].

By fuzzy blending, that is, adopting a center-average defuzzifier, product-fuzzy inference and a singleton fuzzifier, overall fuzzy model is inferred as follows:

$$
\begin{cases}
\dot{x}(t) = \sum_{i=1}^{r} h_i\big(x(t)\big)\Big[A_i\big(r_t\big)x(t) + B_i\big(r_t\big)\big(u(t) + f\big(x(t),t\big)\big)\Big], \\
y(t) = C\big(r_t\big)x(t)
\end{cases}
\tag{5.4}
$$

where $h_i\big(x(t)\big)$ is the membership function given by

$$
h_i\big(x(t)\big) = \frac{\Pi_{j=1}^{n}u_{ij}\big(x_j(t)\big)}{\sum_{i=1}^{r}\Pi_{j=1}^{n}u_{ij}\big(x_j(t)\big)},
$$

in which $\mu_{ij}\big(x_j(t)\big)$ is the grade of membership of $x_j(t)$ in μ_{ij}. For all $t > 0$, it is satisfied that $h_i\big(x(t)\big) \geq 0$ and $\sum_{i=1}^{r} h_i\big(x(t)\big) = 1$. For simplicity, $r(t) = m$ is used for notation in the following.

Assumption 5.1

The unknown continuous-time vector-valued nonlinearity of the system (5.1) is bounded by

$$
\big\| f\big(x(t),t\big)\big\| \leq \alpha + \beta\|y(t)\|,
\tag{5.5}
$$

where α and β are unknown positive constants.

Remark 5.2

With respect to the above nonlinearity, the usual Lipschitz condition or one-sided Lipschitz condition is commonly applied for investigation. In fact, the most general form should be $\big(f\big(x(t),t\big) - F_1 x(t)\big)^{T}\big(f\big(x(t),t\big) - F_2 x(t)\big) \leq 0$, in which F_1 and F_2 are the matrices with appropriate dimensions. This description is quite general that includes the usual Lipschitz condition and norm-bounded condition as special cases. However, for the purpose of adaptive control, we adopted the norm-bounded condition.

5.3 MAIN RESULTS

5.3.1 SLIDING MODE OBSERVER DESIGN

In this section, it is considered that the premise variables are the components of the plant that are not accessible. Then, the following state observer is adopted.

Plant Rule i : **IF** $\hat{x}_1(t)$ is M_{i1} **and** $\hat{x}_2(t)$ is M_{i2} **and** $\cdots$ **and** $\hat{x}_n(t)$ is M_{in}
THEN

$$
\begin{cases}
\dot{\hat{x}}(t) = A_{i,m}\hat{x}(t) + B_{i,m}(u(t) + v_s(t)) + L_{i,m}(y(t) - \hat{y}(t)) \\
\hat{y}(t) = C_m\hat{x}(t) \\
\hat{x}(0) = \hat{\varphi}(0)
\end{cases}
\tag{5.6}
$$

where $\hat{x}(t)$ and $\hat{y}(t)$ are the estimation of the state $x(t)$ and output $y(t)$, respectively. $v_s(t)$ is a compensator selected to attenuate the effect of unknown function $f(x(t),t)$. $L_{i,m}$ is the observer gain with appropriate dimensions to be determined later.

Then, the dynamics fuzzy observer (5.6) is inferred as

$$
\begin{cases}
\dot{\hat{x}}(t) = \sum_{i=1}^{r} h_i(\hat{x}(t))[A_{i,m}\hat{x}(t) + B_{i,m}(u(t) + v_s(t)) + L_{i,m}(y(t) - \hat{y}(t))], \\
\hat{y}(t) = C_m\hat{x}(t)
\end{cases}
\tag{5.7}
$$

Define $e(t) = x(t) - \hat{x}(t)$ as the estimation error. Combining (5.4) and (5.7), the error dynamics can be obtained as follows:

$$
\begin{cases}
\dot{e}(t) = \sum_{i=1}^{r} h_i(\hat{x}(t))[(A_{i,m} - L_{i,m}C_m)e(t) + B_{i,m}(f(x(t),t) - v_s(t))] + w(t), \\
y_e(t) = C_m e(t)
\end{cases}
\tag{5.8}
$$

where $w(t) = \sum_{i=1}^{r}(h_i(x(t)) - h_i(\hat{x}(t)))\dot{x}(t)$ is seen as disturbance.

Remark 5.3

In the error dynamics (5.8), the premise variable in the main part of the fuzzy system is $\hat{x}(t)$, which is in accordance with the observer fuzzy system (5.7). So the following stability analysis will be less conservative from theoretical point of view.

In the following, a distinguished fuzzy integral sliding surface is given. Before moving on, the following general assumption is proposed based on Ref. [21]:
The $n \times m$ matrices $\bar{B}_m$ are defined by

$$\bar{B}_m = \frac{1}{r}\sum_{i=1}^{r} B_{i,m},$$

satisfying the rank constraint that $\text{rank}(\bar{B}_m) = m$, that is, $\bar{B}_m$ have full column rank. Then, we define

$$V_m = \frac{1}{2}[\bar{B}_m - B_{1,m} \quad \bar{B}_m - B_{2,m} \quad \cdots \quad \bar{B}_m - B_{r,m}].$$

$$U(h) = \text{diag}\{1 - 2h_1(\hat{x}(t)), 1 - 2h_2(\hat{x}(t)), \dots, 1 - 2h_r(\hat{x}(t))\}$$

$$W = [I \quad I \quad \cdots \quad I]^T$$

Therefore, we have

$$\bar{B}_m + V_m U(h)W = \bar{B}_m + \frac{1}{2}[(\bar{B}_m - B_{1,m})(1 - 2h_1(\hat{x}(t))) + \cdots$$

$$+ (\bar{B}_m - B_{r,m})(1 - 2h_r(\hat{x}(t)))]$$

$$= \bar{B}_m + \frac{1}{2}\bar{B}_m[(1 - 2h_1(\hat{x}(t))) + \cdots + (1 - 2h_r(\hat{x}(t)))]$$

$$- \frac{1}{2}[B_{1,m}(1 - 2h_1(\hat{x}(t))) + \cdots + B_{r,m}(1 - 2h_r(\hat{x}(t)))]$$

$$= \bar{B}_m + \frac{1}{2}\bar{B}_m(r - 2\sum_{i=1}^{r} h_i(\hat{x}(t))) - \frac{1}{2}\sum_{i=1}^{r} B_{i,m}$$

$$+ \sum_{i=1}^{r} h_i(\hat{x}(t))B_{i,m}$$

$$= \sum_{i=1}^{r} h_i(\hat{x}(t))B_{i,m}$$

Thus, the observer fuzzy system (5.7) is further rewritten as:

$$\begin{cases} \dot{\hat{x}}(t) = \sum_{i=1}^{r} h_i(\hat{x}(t))[A_{i,m}\hat{x}(t) + (\bar{B}_m + \Delta\bar{B}_m)(u(t) \\ \qquad\qquad + v_s(t)) + L_{i,m}(y(t) - \hat{y}(t))], \\ \hat{y}(t) = C_m\hat{x}(t) \end{cases} \qquad (5.9)$$

where $\Delta\bar{B}_m = V_m U(h)W$, in which we have $U^T(h)U(h) \le I_{r\times n}$.

Remark 5.4

More generally, $\bar{B}_m$ can be chosen as the convex combination of $B_{i,m}$, that is, $\bar{B}_m = \sum_{i=1}^{r} \kappa_i B_{i,m}$, in which $\sum_{i=1}^{r} \kappa_i = 1$ with $\kappa_i \geq 0$. On the other hand, if $B_{i,m} = B_{j,m} (i, j = 1, 2, \ldots, r)$, then $V_m = 0$, that is, $\Delta\bar{B}_m = 0$, which is the case that the most existing literature has been considered. Thus, the investigation here is more general since we do not have to assume that $B_{i,m}$ is plant-rule-independent and has full column rank.

5.3.2 DESIGN OF SLIDING SURFACE

In this part, to better accommodate the characteristic of T-S fuzzy systems, based on the state observer (5.6), we propose the following fuzzy integral switching surface function:

$$s(t) = G\hat{x}(t) - \int_0^t \sum_{i=1}^{r} h_i(\hat{x}(s))G(A_{i,m} + \bar{B}_m K_{i,m})\hat{x}(s)ds$$

$$- \int_0^t G\Delta\bar{B}_m(u(s) + v_s(s))ds, \tag{5.10}$$

where $G \in \mathbb{R}^{m \times n}$ is chosen such that $G\bar{B}_m$ is nonsingular, $K_{i,m} \in \mathbb{R}^{m \times n}$ is adopted to stabilize the sliding mode dynamics, so it is selected such that $A_{i,m} + \bar{B}_m K_{i,m}$ is Hurwitz that ensures good stability property.

Based on the systems (5.9) and (5.10), we have

$$\dot{s}(t) = \sum_{i=1}^{r} h_i(\hat{x}(t))G\left[L_{i,m}C_m e(t) - \bar{B}_m K_{i,m}\hat{x}(t) \right] + G\bar{B}_m(u(t) + v_s(t)). \tag{5.11}$$

When the state trajectories of the system (5.9) reached onto the sliding surface, that is, $s(t) = 0$, $\dot{s}(t) = 0$. By $\dot{s}(t) = 0$, we have the equivalent control variable as

$$u_{eq}(t) = \sum_{i=1}^{r} h_i(\hat{x}(t))[K_{i,m}\hat{x}(t) - (G\bar{B}_m)^{-1}GL_{i,m}C_m e(t)] - v_s(t). \tag{5.12}$$

Then, by substituting (5.12) into (5.9), the following sliding mode dynamics can be obtained:

$$\dot{\hat{x}}(t) = \sum_{i=1}^{r} h_i(\hat{x}(t))\left[(A_{i,m} + (\bar{B}_m + \Delta\bar{B}_m)K_{i,m}\hat{x}(t) + \mathcal{I}_m L_{i,m}C_m e(t) \right], \tag{5.13}$$

where $\mathcal{I}_m = \mathcal{I} - (\bar{B}_m + \Delta\bar{B}_m)(G\bar{B}_m)^{-1}G$.

Here, the purpose of this paper is to design observer-based sliding mode controller for the system (1) such that the following two conditions are satisfied:

i. The error dynamical system (5.8) with $w(t) = 0$ and the sliding mode dynamics (5.13) are stochastically stable.
ii. Under zero-initial condition, the following H_∞ performance measurement is satisfied

$$J = \mathbf{E} \int_0^{+\infty} \left[y_e^T(s) y_e(s) - \gamma^2 w^T(s) w(s) \right] ds < 0 \tag{5.14}$$

with γ being a positive constant.

Next, we have to design $v_s(t)$ in advance since the nonlinearity $f(x(t), t)$ is not available. Therefore, we use $\hat{\alpha}(t)$ and $\hat{\beta}(t)$ to estimate the unknown parameters α and β, respectively. The corresponding estimation errors are defined as $\tilde{\alpha}(t) = \hat{\alpha}(t) - \alpha$ and $\tilde{\beta}(t) = \hat{\beta}(t) - \alpha$.

Thus, the compensator $v_s(t)$ with relevant parameters is designed as:

$$v_s(t) = (|\hat{\alpha}(t)| + |\hat{\beta}(t)| \, \| y(t) \| + \varepsilon) sgn(s_e(t)), \tag{5.15}$$

where $s_e(t) = \sum_{i=1}^{r} h_i(\hat{x}(t)) B_{i,m}^T P_m e(t)$, in which it is constrained that $B_{i,m}^T P_m = N_{i,m} C_m$ with P_m and $N_{i,m}$ being determined later. $\varepsilon > 0$ is a small constant. And the adaptive gains are deigned as

$$\dot{\hat{\alpha}}(t) = l_1 \| s_e(t) \|, \dot{\hat{\beta}}(t) = l_2 \| y(t) \| \| s_e(t) \|,$$

with l_1 and l_2 being the positive scalars chosen by designers.

Remark 5.5

Here, the requirement $B_{i,m}^T P_m = N_{i,m} C_m$ is reasonable. By this constraint, it can be seen that

$$s_e(t) = \sum_{i=1}^{r} h_i(\hat{x}(t)) N_{i,m} C_m e(t)$$

$$= \sum_{i=1}^{r} h_i(\hat{x}(t)) N_{i,m} (y(t) - \hat{y}(t)).$$

Thus, the compensator $v_s(t)$ can be realized in the design process based on the estimated states and output measurement variables.

Remark 5.6

Compared to traditional Lesbeuge observer, the proposed sliding mode observer (5.6) has kept some attractive advantages of SMC, such as robustness to disturbances and low sensitivity to the system parameter variations, which is verified in (5.13). On the other hand, besides the asymptotic properties, finite-time convergence property is also an important feature of sliding mode observer schemes.

5.3.3 Stochastic Stability and H-Infinity Performance Analysis

In this section, conditions will be given to ensure stochastic stability of the systems (5.13) and (5.8) with an H_∞ disturbance attenuation level γ.

Theorem 5.1

Given scalar $\gamma > 0$, the error dynamic system (5.8) and the sliding mode dynamics (5.13) are stochastically stable with an H_∞ performance level γ, if there exist positive-definite matrix $P_m > 0$, $Y_{i,m}$ with appropriate dimensions, scalars $\epsilon_m > 0$ and $a_m > 0$ such that the following conditions hold for all $m \in S$

$$a_m I - P_m \leq 0, \qquad (5.16)$$

$$\begin{bmatrix} \Theta^1_{i,m} & 0 & 0 & P_m & P_m V_m & 0 \\ * & \Theta^2_{i,m} & P_m & 0 & 0 & C_m^T Y_{i,m}^T \\ * & * & -\gamma^2 I & 0 & 0 & 0 \\ * & * & * & -a_m \lambda_I^{-1} I & 0 & 0 \\ * & * & * & * & -\epsilon_m I & 0 \\ * & * & * & * & * & -P_m \end{bmatrix} < 0, \qquad (5.17)$$

where

$$\Theta^1_{i,m} = \mathrm{He}\{P_m(A_{i,m} + \bar{B}_m K_{i,m})\} + \epsilon_m K_{i,m}^T W^T W K_{i,m} + \sum_{n=1}^{s} \pi_{mn}(h) P_m,$$

$$\Theta^2_{i,m} = \mathrm{He}\{P_m A_{i,m} - Y_{i,m} C_m\} + C_m^T C_m + \sum_{n=1}^{s} \pi_{mn}(h) P_m,$$

$$\lambda_I = \lambda_{\max}\{\mathcal{I}_m \mathcal{I}_m^T\},$$

and in which $K_{i,m}$ is selected such that $A_{i,m} + \bar{B}_m K_{i,m}$ is Hurwitz. Moreover, the state observer gain is given by $L_{i,m} = P_m^{-1} Y_{i,m}$.

Proof: Consider the Lyapunov functional of the form:

$$V(\hat{x}(t), e(t), r_t) = \hat{x}^T(t)P(r_t)\hat{x}(t) + e^T(t)P(r_t)e(t) + l_1^{-1}\tilde{\alpha}^2(t) + l_2^{-1}\tilde{\beta}^2(t). \quad (5.18)$$

Then, we have

$$\mathcal{L}V(\hat{x}(t), e(t), r_t) = 2\hat{x}^T(t)P_m \sum_{i=1}^{r} h_i(\hat{x}(t))\Big[(A_{i,m} + (\bar{B}_m + \Delta\bar{B}_m)K_{i,m})\hat{x}(t)$$

$$+ \mathcal{I}_m L_{i,m} C_m e(t)\Big] + \hat{x}^T(t)\sum_{n=1}^{s} \pi_{mn}(h)P_n\hat{x}(t)$$

$$+ 2e^T(t)P_m \sum_{i=1}^{r} h_i(\hat{x}(t))[(A_{i,m} - L_{i,m}C_m)e(t)$$

$$+ B_{i,m}(f(x(t),t) - v_s(t))] + e^T(t)\sum_{n=1}^{s} \pi_{mn}(h)P_n e(t)$$

$$+ 2l_1^{-1}\tilde{\alpha}(t)\dot{\tilde{\alpha}}(t) + 2l_2^{-1}\tilde{\beta}(t)\dot{\tilde{\beta}}(t) \quad (5.19)$$

On the other hand, it holds that

$$2\hat{x}^T(t)P_m\Delta\bar{B}_m K_{i,m}\hat{x}(t) \le \epsilon_m^{-1}\hat{x}^T(t)P_m V_m V_m^T P_m\hat{x}(t)$$

$$+ \epsilon_m \hat{x}^T(t)K_{i,m}^T W^T W K_{i,m}\hat{x}(t), \quad (5.20)$$

$$2\hat{x}^T(t)P_m\mathcal{I}_{A_m}L_{i,m}C_m e(t) \le \hat{x}^T(t)P_m\mathcal{I}_m P_m^{-1}\mathcal{I}_m^T P_m\hat{x}(t)$$

$$+ e^T(t)C_m^T L_{i,m}^T P_m L_{i,m}C_m e(t)$$

$$\le a_m^{-1}\lambda_{\mathcal{I}}\hat{x}^T(t)P_m P_m\hat{x}(t)$$

$$+ e^T(t)C_m^T L_{i,m}^T P_m L_{i,m}C_m e(t), \quad (5.21)$$

and by the $v_s(t)$ in (5.15), it follows

$$2e^T(t)\sum_{i=1}^{r} h_i(\hat{x}(t))P_m B_{i,m}(f(x(t),t) - (\mid\hat{\alpha}(t)\mid + \mid\hat{\beta}(t)\parallel\mid y(t)\parallel + \varepsilon)$$

$$sgn(s_e(t))) + 2l_1^{-1}\tilde{\alpha}(t)\dot{\tilde{\alpha}}(t) + 2l_2^{-1}\tilde{\beta}(t)\dot{\tilde{\beta}}(t)$$

$$\le -2\varepsilon \parallel s_e(t)\parallel < 0. \quad (5.22)$$

Overall, we have

$$\mathcal{L}V(\hat{x}(t), e(t), r_t) \le \sum_{i=1}^{r} h_i(\hat{x}(t))\eta^T(t)\hat{\Gamma}_{i,m}\eta(t), \quad (5.23)$$

where $\eta(t) = [x^T(t)\ e^T(t)]^T$, and

$$\hat{\Gamma}_{i,m} = \Gamma_{i,m} + diag\{\sum_{n=1}^{s} \pi_{mn}(h)P_n, \sum_{n=1}^{s} \pi_{mn}(h)P_n\},$$

in which

$$\Gamma_{i,m} = \begin{bmatrix} \Gamma^1_{i,m} & 0 \\ 0 & \Gamma^2_{i,m} \end{bmatrix}$$

with

$$\Gamma^1_{i,m} = \mathrm{Sym}\{P_m(A_{i,m} + \bar{B}_m K_{i,m})\} + \epsilon_m^{-1} P_m V_m V_m^T P_m$$

$$+ \epsilon_m K_{i,m}^T W^T W K_{i,m} + a_m^{-1} \lambda_{\mathcal{I}} P_m P_m,$$

$$\Gamma^2_{i,m} = \mathrm{Sym}\{P_m((A_{i,m} - L_{i,m} C_m)\} + C_m^T L_{i,m}^T P_m L_{i,m} C_m.$$

By letting $P_m L_{i,m} = Y_{i,m}$ and using Schur complement, it is known from (5.17) that $\hat{\Gamma}_{i,m} < 0$. Hence, denote a scalar $\mu \triangleq \lambda_{\min}\{-\hat{\Gamma}_{i,m}\} > 0$ such that

$$\mathcal{L}V(\hat{x}(t), e(t), r_t) \leq -\mu \|\hat{x}(t)\|^2. \tag{5.24}$$

Therefore, by Dynkin's formula, we get for any $t > 0$

$$\mathbf{E}V(\hat{x}(t), e(t), r_t) - \mathbf{E}V(\hat{x}(0), e(0), r_0) \leq -\mu \mathbf{E}\int_0^t \|\hat{x}(s)\|^2\, ds, \tag{5.25}$$

which yields

$$\mathbf{E}\int_0^t \|\hat{x}(s)\|^2\, ds \leq \mu^{-1}\mathbf{E}V(\hat{x}(0), e(0), r_0). \tag{5.26}$$

Then, it is easy to obtain that the sliding mode dynamics (5.13) is stochastically stable in the case of $w(t) = 0$ according to Definition 1.3. The proof of error dynamical system (5.8) follows the same way.

Next, we consider the H_∞ performance of the overall closed-loop system. Under the zero-initial condition, $\mathbf{E}V(t) = \mathbf{E}\int_0^{+\infty} \mathcal{L}V(s)ds \geq 0$. Thus,

$$J = E\int_0^{+\infty}\left[y_e^T(s)y(s) - \gamma^2 w^T(s)w(s)\right]ds$$

$$\leq E\int_0^{+\infty}\left[y_e^T(s)y(s) - \gamma^2 w^T(s)w(s) + \mathcal{L}V(s)\right]ds, \tag{5.27}$$

$$= E\int_0^{+\infty}\sum_{i=1}^{r} h_i(\hat{x}(s))\zeta^T(s)\bar{\Gamma}_{i,m}\zeta(s)ds$$

where $\zeta(t) = [\hat{x}^T(t) \; e^T(t) \; w^T(t)]^T$, and

$$\overline{\Gamma}_{i,m} = \breve{\Gamma}_{i,m} + diag\{\sum_{n=1}^{s} \pi_{mn}(h)P_n, \sum_{n=1}^{s} \pi_{mn}(h)P_n, 0\},$$

with

$$\breve{\Gamma}_{i,m} = \begin{bmatrix} \Gamma_{i,m} + \begin{bmatrix} 0 & 0 \\ 0 & C_m^T C_m \end{bmatrix} & \begin{bmatrix} 0 \\ P_m \end{bmatrix} \\ * & -\gamma^2 I \end{bmatrix}.$$

By utilizing Schur complement and (5.17), it is easily known that $\overline{\Gamma}_{i,m} < 0$ means $J < 0$. Thus, the sliding mode dynamics (5.13) with error dynamics (5.8) is stochastically stable with a H_∞ disturbance attenuation level γ. This completes the proof.

In the above theorem, conditions for the stochastic stability with H_∞ performance level γ have been given for the sliding mode dynamics. As mentioned before, the stability analysis with respect to the sliding mode observer and the error dynamics is in accordance with the estimated $\hat{x}(t)$ in the membership functions, which avoid the errors from plant state $x(t)$. Therefore, the results will be less conservative.

Remark 5.7

For MJSs, the proof ends in Theorem 5.1, since the jump time of a Markov chain is, in general, exponentially distributed, and the results obtained for the MJS are intrinsically conservative due to constant TRs. However, the S-MJSs, due to their relaxed conditions on the probability distributions, have much broader applications than the conventional MJS, which also cause the LMI conditions in Theorem 5.1 are not solvable since they are not linear. Thus, a novel way is provided in the following to deal with this problem.

Theorem 5.2

Given scalar $\gamma > 0$, the error dynamic system (5.8) and the sliding mode dynamics (5.13) are stochastically stable with an H_∞ performance level γ, if there exist positive-definite matrix $P_m > 0$, $T_{mn} > 0$, $Z_{mn} > 0$, $Y_{i,m}$ with appropriate dimensions, scalars $\epsilon_m > 0$ and $a_m > 0$ such that (5.16) holds and the following conditions hold for all $m \in \mathcal{S}$
If $m \in I_{m,k}, \forall l \in I_{m,uk}, I_{m,k} \triangleq \{k_{m,1}, k_{m,2}, \ldots, k_{m,o_1}\}$,

$$\begin{bmatrix} \mathcal{A}_{i,m}^{11} & 0 & 0 & \mathcal{A}_{i,m}^{12} & \mathcal{A}_{i,m}^{13} & 0 \\ * & \mathcal{A}_{i,m}^{14} & P_m & \mathcal{A}_{i,m}^{15} & 0 & \mathcal{A}_{i,m}^{16} \\ * & * & -\gamma^2 I & 0 & 0 & 0 \\ * & * & * & \mathcal{A}_m^{17} & 0 & 0 \\ * & * & * & * & \mathcal{A}_m^{18} & 0 \\ * & * & * & * & * & \mathcal{A}_m^{19} \end{bmatrix} < 0, \qquad (5.28)$$

If $m \in I_{m,uk}$, $\forall l \in I_{m,uk}$, $I_{m,k} \triangleq \{k_{m,1}, k_{m,2}, \ldots, k_{m,o2}\}$, $l \neq m$,

$$P_m - P_l \geq 0, \tag{5.29}$$

$$\begin{bmatrix} \mathcal{A}_{i,m}^{21} & 0 & 0 & \mathcal{A}_{i,m}^{22} & \mathcal{A}_{i,m}^{23} & 0 \\ * & \mathcal{A}_{i,m}^{24} & P_m & \mathcal{A}_{i,m}^{25} & 0 & \mathcal{A}_{i,m}^{26} \\ * & * & -\gamma^2 I & 0 & 0 & 0 \\ * & * & * & \mathcal{A}_m^{27} & 0 & 0 \\ * & * & * & * & \mathcal{A}_m^{28} & 0 \\ * & * & * & * & * & \mathcal{A}_m^{29} \end{bmatrix} < 0, \tag{5.30}$$

where

$$\mathcal{A}_{i,m}^{11} = \text{He}\{P_m(A_{i,m} + \bar{B}_m K_{i,m})\} + \epsilon_m K_{i,m}^T W^T W K_{i,m}$$

$$+ \sum_{n \in I_{m,k}} \left[\frac{(\lambda_{mn})^2}{4} T_{mn} + \pi_{mn}(P_n - P_l) \right],$$

$$\mathcal{A}_{i,m}^{12} = [P_m \quad P_m V_m \quad 0],$$

$$\mathcal{A}_m^{13} = [(P_{k_{m,1}} - P_l) \quad \ldots \quad (P_{k_{m,o1}} - P_l)],$$

$$\mathcal{A}_{i,m}^{14} = \text{He}\{P_m A_{i,m} - Y_{i,m} C_m\} + C_m^T C_m$$

$$+ \sum_{n \in I_{m,k}} \left[\frac{(\lambda_{mn})^2}{4} T_{mn} + \pi_{mn}(P_n - P_l) \right],$$

$$\mathcal{A}_{i,m}^{15} = [0 \quad 0 \quad C_m^T Y_{i,m}^T],$$

$$\mathcal{A}_m^{16} = \mathcal{A}_m^{13},$$

$$\mathcal{A}_{i,m}^{17} = diag\{-a\lambda_I^{-1}I, -\epsilon_m I, -P_m\},$$

$$\mathcal{A}_m^{18} = \mathcal{A}_m^{19} = [-T_{mk_{m,1}} \quad \ldots \quad -T_{mk_{m,o1}}],$$

$$\mathcal{A}_{i,m}^{21} = \text{He}\{P_m(A_{i,m} + \bar{B}_m K_{i,m})\} + \epsilon_m K_{i,m}^T W^T W K_{i,m}$$

$$+ \sum_{n \in I_{m,k}} \left[\frac{(\lambda_{mn})^2}{4} Z_{mn} + \pi_{mn}(P_n - P_l) \right],$$

$$\mathcal{A}_m^{23} = [(P_{k_{m,1}} - P_l) \ \cdots \ (P_{k_{m,o2}} - P_l)],$$

$$\mathcal{A}_{i,m}^{24} = \mathrm{He}\{P_m A_{i,m} - Y_{i,m} C_m\} + C_m^T C_m$$

$$+ \sum_{n \in I_{m,k}} \left[\frac{(\lambda_{mn})^2}{4} Z_{mn} + \pi_{mn}(P_n - P_l) \right],$$

$$\mathcal{A}_m^{28} = \mathcal{A}_m^{29} = [-Z_{mk_{m,1}} \ \cdots \ -Z_{mk_{m,o2}}],$$

$$\mathcal{A}_{i,m}^{22} = \mathcal{A}_{i,m}^{12}, \mathcal{A}_{i,m}^{25} = \mathcal{A}_{i,m}^{15}, \mathcal{A}_m^{26} = \mathcal{A}_m^{23}, \mathcal{A}_{i,m}^{27} = \mathcal{A}_{i,m}^{17}.$$

Moreover, the state observer gain is given by $L_{i,m} = P_m^{-1} Y_{i,m}$.

Proof: Let us consider the following two cases:

Case I: $m \in I_{m,k}$

First, denote $\lambda_{m,k} \triangleq \sum_{n \in I_{m,k}} \pi_{mn}(h)$. Since $I_{m,uk} \neq \varnothing$, it holds that $\lambda_{m,k} < 0$. Note that

$\sum_{n=1}^{s} \pi_{mn}(h) P_n$ can be represented as

$$\sum_{n=1}^{s} \pi_{mn}(h) P_n = \left(\sum_{n \in I_{m,k}} + \sum_{n} \in I_{m,uk} \right) \pi_{mn}(h) P_n$$

$$= \sum_{n \in I_{m,k}} \pi_{mn}(h) P_n - \lambda_{m,k} \sum_{n \in I_{m,uk}} \frac{\pi_{mn}(h)}{-\lambda_{m,k}} P_n, \tag{5.31}$$

and it is obvious that $0 \leq \pi_{mn}(h)/-\lambda_{m,k} \leq 1 \ (n \in I_{m,uk})$ and $\sum_{n \in I_{m,uk}} \frac{\pi_{mn}(h)}{-\lambda_{m,k}} = 1$. So for $\forall l \in I_{m,uk}$, there is

$$\bar{\Gamma}_{i,m} = \sum_{n \in I_{m,uk}} \frac{\pi_{mn}(h)}{-\lambda_{m,k}} \left[\check{\Gamma}_{i,m} + diag\{ \sum_{n \in I_{m,k}} \pi_{mn}(h)(P_n - P_l), \right.$$

$$\left. \sum_{n \in I_{m,k}} \pi_{mn}(h)(P_n - P_l), 0 \} \right] \tag{5.32}$$

Therefore, for $0 \leq \pi_{mn}(h) \leq -\lambda_{m,k}, \bar{\Gamma}_{i,m} < 0$ is equivalent to

$$\check{\Gamma}_{i,m} + diag\{ \sum_{n \in I_{m,k}} \pi_{mn}(h)(P_n - P_l), \sum_{n \in I_{m,k}} \pi_{mn}(h)(P_n - P_l), 0 \} < 0. \tag{5.33}$$

In formula (5.33), it is true that

$$\sum_{n\in I_{m,k}} \pi_{mn}(h)(P_n - P_l) = \sum_{n\in I_{m,k}} \pi_{mn}(P_n - P_l) + \sum_{n\in I_{m,k}} \Delta\pi_{mn}(h)(P_n - P_l). \quad (5.34)$$

Then, by virtue of Lemma 1.4 and for any $T_{mn} > 0$, it follows that

$$\sum_{n\in I_{m,k}} \Delta\pi_{mn}(h)(P_n - P_l) = \sum_{n\in I_{m,k}} \left[\frac{1}{2} \Delta\pi_{mn}(h)((P_n - P_l) + (P_n - P_l)) \right]$$

$$\leq \sum_{n\in I_{m,k}} \left[\frac{(\lambda_{mn})^2}{4} T_{mn} + (P_n - P_l)(T_{mn})^{-1}(P_n - P_l)^T \right]. \quad (5.35)$$

Combining (5.31)–(5.35), then by applying Schur complement, we can see that (5.28) guarantees $\overline{\Gamma}_{i,m} < 0$ when $m \in I_{m,k}$.

Case II: $m \in I_{m,uk}$.

Similarly, denote $\lambda_{m,k} \triangleq \sum_{n\in I_{m,k}} \pi_{mn}(h)$. Since $I_{m,k} \neq \varnothing$, it holds that $\lambda_{m,k} > 0$. Now $\sum_{n=1}^{s} \pi_{mn}(h)P_n$ can be represented as

$$\sum_{n=1}^{s} \pi_{mn}(h)P_n = \sum_{n\in I_{m,k}} \pi_{mn}(h)P_n + \pi_{mm}(h)P_m + \sum_{n\in I_{m,uk}, n\neq m} \pi_{mn}(h)P_n$$

$$= \sum_{n\in I_{m,k}} \pi_{mn}(h)P_n + \pi_{mm}(h)P_m$$

$$- (\pi_{mm}(h) + \lambda_{m,k}) \sum_{n\in I_{m,uk}, n\neq m} \frac{\pi_{mn}(h)P_n}{-\pi_{mm}(h) - \lambda_{m,k}}, \quad (5.36)$$

and it is obvious that $0 \leq \pi_{mn}(h)/-\pi_{mm}(h) - \lambda_{m,k} \leq 1$ $(n \in I_{m,uk})$ and $\sum_{n\in I_{m,uk}, n\neq m} \frac{\pi_{mn}(h)}{-\pi_{mm}(h) - \lambda_{m,k}} = 1$. So for $\forall l \in I_{m,uk}, l \neq m$,

$$\overline{\Gamma}_{i,m} = \sum_{n\in I_{m,uk}, n\neq m} \frac{\pi_{mn}(h)}{-\pi_{mm}(h) - \lambda_{m,k}} \left[\check{\Gamma}_{i,m} + diag\{\Lambda_m, \Lambda_m, 0\} \right], \quad (5.37)$$

where $\Lambda_m = \pi_{mm}(h)(P_m - P_l) + \sum_{n\in I_{m,k}} \pi_{mn}(h)(P_n - P_l)$.

Therefore, for $0 \leq \pi_{mn}(h) \leq -\pi_{mm}(h) - \lambda_{m,k}$, $\hat{\Gamma}_{i,m} < 0$ is equivalent to

$$\check{\Gamma}_{i,m} + diag\{\Lambda_m, \Lambda_m, 0\} < 0. \quad (5.38)$$

Since $\pi_{mm}(h) < 0$, (5.38) holds if we have

$$\begin{cases} P_m - P_l \geq 0, \\ \check{\Gamma}_{i,m} + diag\{ \sum_{n \in I_{m,k}} \pi_{mn}(h)(P_n - P_l), \sum_{n \in I_{m,k}} \pi_{mn}(h)(P_n - P_l), 0 \} < 0. \end{cases} \quad (5.39)$$

Also, as in (5.34) and (5.35), for any $Z_{mn} > 0$, we have

$$\sum_{n \in I_{m,k}} \pi_{mn}(h)(P_n - P_l) \leq \sum_{n \in I_{m,k}} \pi_{mn}(P_n - P_l)$$

$$+ \sum_{n \in I_{m,k}} \left[\frac{(\lambda_{mn})^2}{4} Z_{mn} + (P_n - P_l)(Z_{mn})^{-1}(P_n - P_l)^T \right]. \quad (5.40)$$

Combining (5.36)–(5.40), we know that (5.29) and (5.30) guarantee $\overline{\Gamma}_{i,m} < 0$ by applying Schur complement when $m \in I_{m,uk}$. In summary, the systems achieved stochastic stability with an H_∞ performance level γ despite existence of generally uncertain TRs from the above analysis. This completes the proof.

As discussed in many other works, if $B_{i,m} \equiv B_{j,m}$ $(i, j = 1, 2, \ldots, r)$, that is, $\Delta \overline{B}_m = 0$, this case is also possibly existing in practical systems. Therefore, we can conclude the following theorem.

Theorem 5.3

Given scalar $\gamma > 0$, the error dynamic system (5.8) and the sliding mode dynamics (5.13) are stochastically stable with an H_∞ performance level γ, if there exist positive-definite matrix $P_m > 0$, $T_{mn} > 0$, $Z_{mn} > 0$, $Y_{i,m}$ with appropriate dimensions, scalars $\epsilon_m > 0$ and $a_m > 0$ such that (5.16) with the following conditions hold for all $m \in S$

If $m \in I_{m,k}, \forall l \in I_{m,uk}, I_{m,k} \triangleq \{k_{m,1}, k_{m,2}, \ldots, k_{m,o_1}\}$,

$$\begin{bmatrix} \overline{\mathcal{A}}_{i,m}^{11} & 0 & 0 & \overline{\mathcal{A}}_{i,m}^{12} & \mathcal{A}_{i,m}^{13} & 0 \\ * & \overline{\mathcal{A}}_{i,m}^{14} & P_m & \overline{\mathcal{A}}_{i,m}^{15} & 0 & \mathcal{A}_{i,m}^{16} \\ * & * & -\gamma^2 I & 0 & 0 & 0 \\ * & * & * & \overline{\mathcal{A}}_m^{17} & 0 & 0 \\ * & * & * & * & \mathcal{A}_m^{18} & 0 \\ * & * & * & * & * & \mathcal{A}_m^{19} \end{bmatrix} < 0, \quad (5.41)$$

If $m \in I_{m,uk}, \forall l \in I_{m,uk}, I_{m,k} \triangleq \{k_{m,1}, k_{m,2}, \ldots, k_{m,o_2}\}, l \neq m$,

$$P_m - P_l \geq 0 P_m - P_l \geq 0, \quad (5.42)$$

$$\begin{bmatrix} \overline{\mathcal{A}}_{i,m}^{21} & 0 & 0 & \overline{\mathcal{A}}_{i,m}^{22} & \mathcal{A}_{i,m}^{23} & 0 \\ * & \overline{\mathcal{A}}_{i,m}^{24} & P_m & \overline{\mathcal{A}}_{i,m}^{25} & 0 & \mathcal{A}_{i,m}^{26} \\ * & * & -\gamma^2 I & 0 & 0 & 0 \\ * & * & * & \overline{\mathcal{A}}_{m}^{27} & 0 & 0 \\ * & * & * & * & \mathcal{A}_{m}^{28} & 0 \\ * & * & * & * & * & \mathcal{A}_{m}^{29} \end{bmatrix} < 0, \qquad (5.43)$$

where

$$\overline{\mathcal{A}}_{i,m}^{11} = \mathrm{He}\{P_m(A_{i,m} + B_m K_{i,m})\} + \sum_{n \in I_{m,k}} \left[\frac{(\lambda_{mn})^2}{4} T_{mn} + \pi_{mn}(P_n - P_l) \right],$$

$$\overline{\mathcal{A}}_{i,m}^{12} = \overline{\mathcal{A}}_{i,m}^{22} = [P_m \mathcal{I}_m' \ 0],$$

$$\overline{\mathcal{A}}_{i,m}^{14} = \mathrm{He}\{P_m A_{i,m} - Y_{i,m} C_m)\} + C_m^T C_m + \sum_{n \in I_{m,k}} \left[\frac{(\lambda_{mn})^2}{4} T_{mn} + \pi_{mn}(P_n - P_l) \right],$$

$$\overline{\mathcal{A}}_{i,m}^{15} = \overline{\mathcal{A}}_{i,m}^{25} = \begin{bmatrix} 0 & C_m^T Y_{i,m}^T \end{bmatrix},$$

$$\overline{\mathcal{A}}_{i,m}^{17} = \overline{\mathcal{A}}_{i,m}^{27} = diag\{-P_m, -P_m\}$$

$$\overline{\mathcal{A}}_{i,m}^{21} = \mathrm{He}\{P_m(A_{i,m} + B_m K_{i,m})\} + \sum_{n \in I_{m,k}} \left[\frac{(\lambda_{mn})^2}{4} Z_{mn} + \pi_{mn}(P_n - P_l) \right],$$

$$\overline{\mathcal{A}}_{i,m}^{24} = \mathrm{He}\{P_m A_{i,m} - Y_{i,m} C_m\} + C_m^T C_m + \sum_{n \in I_{m,k}} \left[\frac{(\lambda_{mn})^2}{4} Z_{mn} + \pi_{mn}(P_n - P_l) \right],$$

with $\mathcal{I}_m' = I - B_m(GB_m)^{-1}G$, and the other notations are defined as in Theorem 5.2. Moreover, the state observer gain is given by $L_{i,m} = P_m^{-1} Y_{i,m}$.

Remark 5.8

From Theorem 5.2 and Theorem 5.3, we give stochastic stability with H_∞ performance analysis of sliding mode dynamics under two conditions: the input matrices are plant-rule-dependent or independent. The results show the analysis and synthesis of the fuzzy control scheme will be more simple when the input matrices are plant-rule-independent as in most literature. Therefore, the result of Theorem 5.2 is valuable since a potential effective way to check the synthesis of T-S fuzzy systems with plant-rule-dependent input matrices has been established.

5.3.4 REACHABILITY ANALYSIS

In this section, we deal with reachability of the sliding surface $s(t) = 0$; it will be confirmed that the proposed control scheme will drive the estimated state onto the predesigned sliding surface in a finite-time interval.

Theorem 5.4

Suppose that the sliding surface function is proposed in (5.10), and the conditions in Theorem 5.2 are solvable. Then, the state trajectories of the observer system (5.6) will be driven onto the sliding surface $s(t) = 0$ in finite time by the fuzzy SMC law synthesized as follows:

$$u(t) = \sum_{i=1}^{r} h_i(\hat{x}(t)) K_{i,m} \hat{x}(t) - v_s(t) - (G\bar{B}_m)^{-1}(\rho(t) + \delta) sgn(s(t)), \qquad (5.44)$$

where δ is a small-positive tuning scalar, and

$$\rho(t) = \max_{m \in S} \sum_{i=1}^{r} h_i(\hat{x}(t))[\| GL_{i,m} \| \| y(t) \| + \| GL_{i,m} C_m \| \| \hat{x}(t) \|].$$

Proof: Choose the following Lyapunov function:

$$V(t) = \frac{1}{2} s^T(t) s(t). \qquad (5.45)$$

Then, the infinitesimal generator of Lyapunov function $V(t)$ along the trajectory of sliding mode dynamics is given as follows:

$$\mathcal{L}V(t) = s^T(t)\dot{s}(t)$$

$$= s^T(t) \sum_{i=1}^{r} h_i(\hat{x}(t)) G\left[L_{i,m} C_m e(t) - \bar{B}_m K_{i,m} \hat{x}(t) \right]$$

$$+ G\bar{B}_m(u(t) + v_s(t)).$$

$$\leq |s(t)| \sum_{i}^{r} h_i(\hat{x}(t))\left[\|GL_{i,m}\| \|y(t)\| + \|GL_{i,m}\| \|\hat{x}(t)\| \right]$$

$$- s^T(t) \sum_{i=1}^{r} h_i(\hat{x}(t)) G\bar{B}_m K_{i,m} \hat{x}(t)$$

$$+ s^T(t) G\bar{B}_m(u(t) + v_s(t))$$

$$\mathcal{L}V(t) = s^T(t)\dot{s}(t)$$

$$= s^T(t) \sum_{i=1}^{r} h_i(\hat{x}(t)) G \left[L_{i,m} C_m e(t) - \bar{B}_m K_{i,m} \hat{x}(t) \right]$$

$$+ G\bar{B}_m (u(t) + v_s(t)).$$

$$\leq |s(t)| \sum_{i} h_i(\hat{x}(t))[\| GL_{i,m} \| \| y(t) \| + \| GL_{i,m} C_m \| \| \hat{x}(t) \|]$$

$$- s^T(t) \sum_{i=1}^{r} h_i(\hat{x}(t)) G\bar{B}_m K_{i,m} \hat{x}(t)$$

$$+ s^T(t) G\bar{B}_m (u(t) + v_s(t)) \tag{5.46}$$

By substituting (5.44) into (5.46), we have

$$\mathcal{L}V(t) \leq -\delta \| s(t) \| < 0 \quad \text{for} \quad s(t) \neq 0. \tag{5.47}$$

Therefore, seeing from (5.47), the reachability of the sliding surface $s(t) = 0$ can be ensured in finite time. This completes the proof.

Remark 5.9

Note that in the synthesis of compensator $v_s(t)$, it is required that $B_{i,m}^T P_m = N_{i,m} C_m$. This equality cannot be solved by applying LMI Tool-box in MatLab environment. To deal with this computational problem, the following equivalent transformation is introduced. Since $B_{i,m}^T P_m = N_{i,m} C_m$, it holds

$$Trace\{ (B_{i,m}^T P_m - N_{i,m} C_m)(B_{i,m}^T P_m - N_{i,m} C_m)^T \} = 0. \tag{5.48}$$

Thus, there exists a positive scalar α_1 such that

$$(B_{i,m}^T P_m - N_{i,m} C_m)(B_{i,m}^T P_m - N_{i,m} C_m)^T < \alpha_1 I. \tag{5.49}$$

By Schur complement, the above inequality is equivalent to

$$\begin{bmatrix} -\alpha_1 I & B_{i,m}^T P_m - N_{i,m} C_m \\ * & -I \end{bmatrix} < 0. \tag{5.50}$$

Hence, the H_∞ performance index γ can be optimized by finding a global solution of the following minimization problem:

$$\min \alpha_1 \text{ in.} \tag{5.51}$$

Subject to conditions in Theorem 5.2 or Theorem 5.3.

Remark 5.10

When $\Delta \bar{B}_m = 0$, the sliding surface function will be correspondingly changed, but the sliding mode dynamics remains the same. So it can be easily seen the observer-based adaptive sliding mode controller (5.54) is also valid for stabilization of the original T-S fuzzy system (5.1) with plant-rule-independent input matrices. Therefore, the provided SMC scheme is universal for both cases.

5.4 NUMERICAL EXAMPLE

In this section, a numerical example is presented to demonstrate the effectiveness of the above results. Consider a single-link robot arm model proposed in Refs. [22,23], where the dynamic equation is given by

$$\ddot{\theta}(t) = -\frac{MgL}{J}\sin(\theta(t)) - \frac{D(t)}{J}\dot{\theta}(t) + \frac{1}{J}u(t),$$

where $\theta(t)$ is the angle position of the arm, $u(t)$ is the control input, M is the mass of the payload, J is the moment of inertia, g is the acceleration of gravity, L is the length of the arm, and $D(t)$ is the coefficient of viscous friction. The values of parameters g and L are given by $g = 9.81$ and $L = 0.5$. It is assumed that the parameter $D(t) = D_0 = 2$ is time-invariant, and the parameters M and J have three different modes as shown in Table 5.1.

The transition-probability-rate matrix that relates the three operation modes is given as follows:

$$\begin{bmatrix} -1.0 + \Delta\pi_{11}(h) & ? & ? \\ ? & ? & 0.5 + \Delta\pi_{23}(h) \\ 0.5 + \Delta\pi_{31}(h) & ? & -1.0 + \Delta\pi_{33}(h) \end{bmatrix}.$$

Let $x_1(t) = \theta(t)$ and $x_2(t) = \dot{\theta}(t)$. Under uncertain conditions [24], the nonlinear term $\sin(x_1(t))$ can be represented as

$$\sin(x_1(t)) = h_1(x_1(t))x_1(t) + \beta h_2(x_1(t))x_1(t),$$

TABLE 5.1

Modes and Values of the Parameters M and J

Mode m	Parameter M	Parameter J
1	1	1
2	1.5	2
3	2	2.5

where $\beta = 0.01 / \pi$, $h_1(x_1(t))$, $h_2(x_1(t)) \in [0,1]$ and $h_1(x_1(t)) + h_2(x_1(t)) = 1$. Thus, the membership functions $h_1(x_1(t))$ and $h_2(x_1(t))$ are given correspondingly as:

$$h_1(x_1(t)) = \begin{cases} \dfrac{\sin(x_1(t)) - \beta x_1(t)}{x_1(t)(1 - \beta)} & x_1(t) \neq 0 \\ 1, & x_1(t) = 0 \end{cases}$$

$$h_2(x_1(t)) = \begin{cases} \dfrac{x_1(t) - \sin(x_1(t))}{x_1(t)(1 - \beta)} & x_1(t) \neq 0 \\ 0, & x_1(t) = 0 \end{cases}.$$

It is evident from the above membership functions that with $x_1(t) = 0$ rad, then $h_1(x_1(t)) = 1$, $h_2(x_1(t)) = 0$ and with $x_1(t) = \pi$ rad or $x_1(t) = -\pi$ rad, then $h_1(x_1(t)) = 0$, $h_2(x_1(t)) = 1$. Thus, if we take nonlinear perturbations into consideration, the state-space representation of single-link robot arm can be expressed by the following two-rule T-S fuzzy system:

Plant Rule 1: IF $x_1(t)$ is "about 0 rad,"
THEN

$$\begin{cases} \dot{x}(t) = A_{1,m}x(t) + B_{1,m}(u(t) + f(x(t),t)) \\ y(t) = C_m x(t). \end{cases}$$

Plant Rule 2: IF $x_1(t)$ is "about π rad or $-\pi$ rad,"
THEN

$$\begin{cases} \dot{x}(t) = A_{2,m}x(t) + B_{2,m}(u(t) + f(x(t),t)) \\ y(t) = C_m x(t). \end{cases}$$

where $x(t) = [x_1^T(t) \ x_2^T(t)]^T$, and

$$A_{1,1} = \begin{bmatrix} 0 & 1 \\ -gL & -D_0 \end{bmatrix}, B_{1,1} = B_{2,1} = \begin{bmatrix} 0 \\ 1 \end{bmatrix},$$

$$A_{1,2} = \begin{bmatrix} 0 & 1 \\ -0.75gL & -0.5D_0 \end{bmatrix}, B_{1,2} = B_{2,2} = \begin{bmatrix} 0 \\ 0.5 \end{bmatrix},$$

$$A_{1,3} = \begin{bmatrix} 0 & 1 \\ -0.8gL & -0.4D_0 \end{bmatrix}, B_{1,3} = B_{2,3} = \begin{bmatrix} 0 \\ 0.4 \end{bmatrix},$$

$$A_{2,1} = \begin{bmatrix} 0 & 1 \\ -\beta gL & -D_0 \end{bmatrix}, C_1 = [0.1 \ 0.3],$$

$$A_{2,2} = \begin{bmatrix} 0 & 1 \\ -0.75\beta gL & -0.5D_0 \end{bmatrix}, C_2 = [-0.3 \ 0.4],$$

$$A_{2,3} = \begin{bmatrix} 0 & 1 \\ -0.8\beta gL & -0.4D_0 \end{bmatrix}, C_3 = [0.2 \ -0.1].$$

In this model, the input matrices are plant-rule-independent, so we can use the conditions in Theorem 5.3 to check the effectiveness of the proposed SMC theory. Let $G = [0 \ 1]$ such that GB_m is nonsingular, $K_{1,1} = [-5 \ -3]$, $K_{1,2} = [-3 \ -2]$, $K_{1,3} = [\ -4 \ \ \ -2 \ ]$, $K_{2,1} = [-3 \ -1]$, $K_{2,2} = [-6 \ -3]$, $K_{2,3} = [-7 \ -6]$, and $\Delta\pi_{mn}(h) \le \lambda_{mn} = |0.1 * \pi_{mn}|$. By solving the optimal minimum problems in Remark 5.9 with $\gamma = 2.5$, we can obtain the following feasible solutions:

$$P_1 = \begin{bmatrix} 0.2692 & 0.0659 \\ 0.0659 & 0.0560 \end{bmatrix}, P_2 = \begin{bmatrix} 0.2761 & 0.0839 \\ 0.0839 & 0.1027 \end{bmatrix},$$

$$P_3 = \begin{bmatrix} 0.2864 & 0.0773 \\ 0.0773 & 0.0910 \end{bmatrix}, T_{1,1} = \begin{bmatrix} 0.3737 & -0.3479 \\ -0.3479 & 11.3044 \end{bmatrix},$$

$$T_{3,1} = \begin{bmatrix} 0.2750 & 0.7166 \\ 0.7166 & 1.8677 \end{bmatrix}, T_{3,3} = \begin{bmatrix} 0.2478 & 0.0124 \\ 0.0124 & 0.2639 \end{bmatrix},$$

$$Z_{2,3} = \begin{bmatrix} 0.6867 & 0.4529 \\ 0.4529 & 1.4028 \end{bmatrix}, Y_{1,1} = \begin{bmatrix} 1.0664 \\ -0.7208 \end{bmatrix},$$

$$Y_{1,2} = \begin{bmatrix} 0.0565 \\ 0.4630 \end{bmatrix}, Y_{1,3} = \begin{bmatrix} 0.1082 \\ 0.7312 \end{bmatrix}, Y_{2,1} = \begin{bmatrix} 3.4828 \\ 1.3304 \end{bmatrix},$$

$$Y_{2,2} = \begin{bmatrix} 0.9349 \\ 0.5850 \end{bmatrix}, Y_{2,3} = \begin{bmatrix} 2.2671 \\ 1.3137 \end{bmatrix}.$$

Thus, we can get the observer gains

$$L_{1,1} = \begin{bmatrix} 10 \\ -24.6565 \end{bmatrix}, L_{1,2} = \begin{bmatrix} -1.5494 \\ 5.7747 \end{bmatrix},$$

$$L_{1,3} = \begin{bmatrix} -2.3201 \\ 10 \end{bmatrix}, L_{2,1} = \begin{bmatrix} 10 \\ 11.9853 \end{bmatrix},$$

$$L_{2,2} = \begin{bmatrix} 2.2015 \\ 3.8993 \end{bmatrix}, L_{2,3} = \begin{bmatrix} 5.2178 \\ 10 \end{bmatrix}.$$

Given the initial conditions $x(0) = [0.5\pi \quad -1]^T$ and $\hat{x}(0) = [0.25\pi \quad -0.5]^T$. The adjustable parameters $\delta = \varepsilon = 0.01$ and $f(x(t)) = 0.1\sin(x_1(t))$. The adaptive laws in (5.15) with parameters $l_1 = l_2 = 1$. Also, in order to reduce the chattering effect of switching signals, $\mathrm{sgn}(s(t))$ is replaced by $\dfrac{s(t)}{\|s(t)\| + 0.01}$. The simulation results are presented in Figures 5.1–5.6. Figure 5.1 gives the membership functions. Figure 5.3 draws the state response of the system (5.1) and the observer system (5.6) under one possible jumping modes. Figure 5.4 depicts the response of sliding surface, and Figure 5.5 shows the control input. Figure 5.6 gives the estimated values $\hat{\alpha}(t)$ and $\hat{\beta}(t)$. Figure 5.7 shows the value of integral item during the whole phase, which includes the reaching phase and the sliding motion phase. What is worthy pointing out is that the H_∞

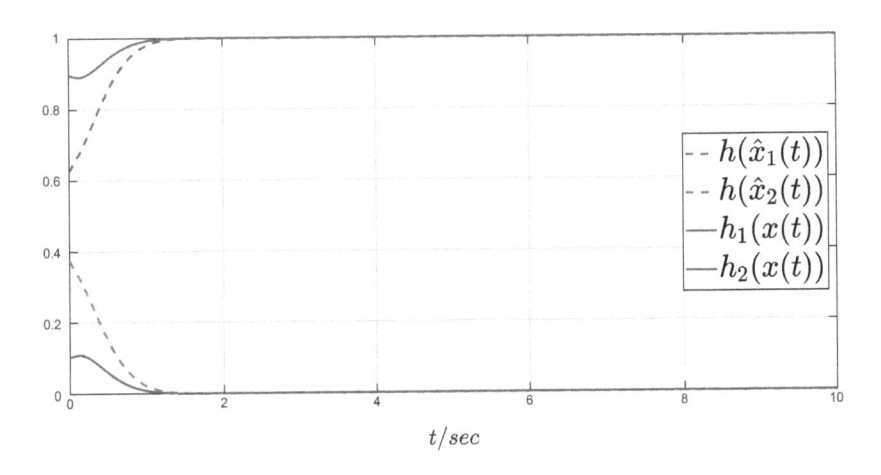

FIGURE 5.1 Membership functions.

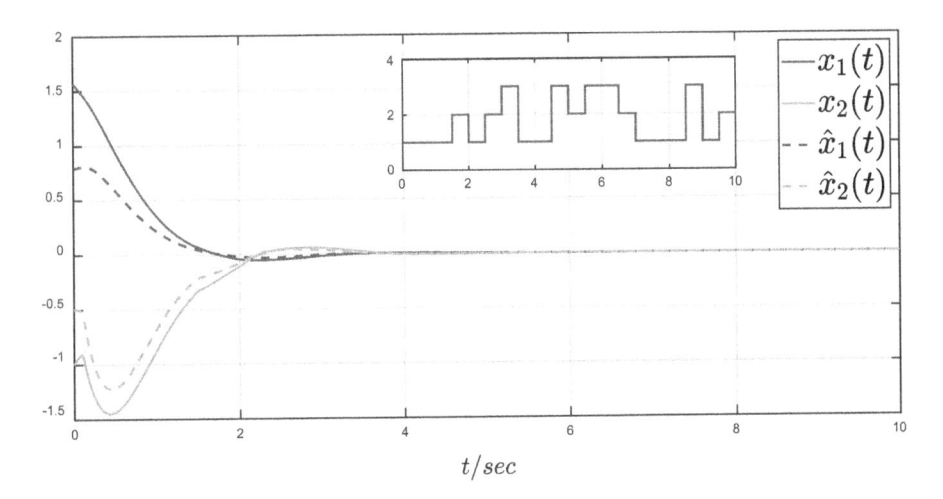

FIGURE 5.2 System state response of $x(t)$ and $\hat{x}(t)$.

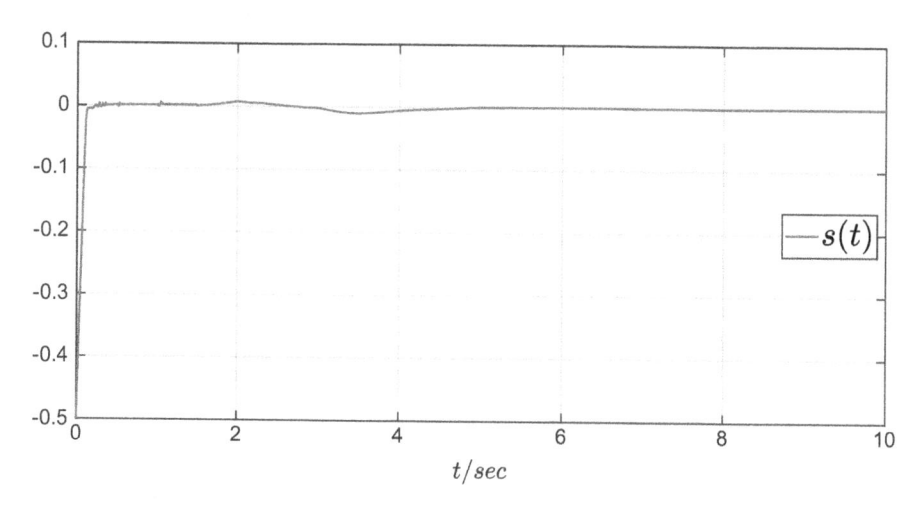

FIGURE 5.3 Sliding surface function $s(t)$.

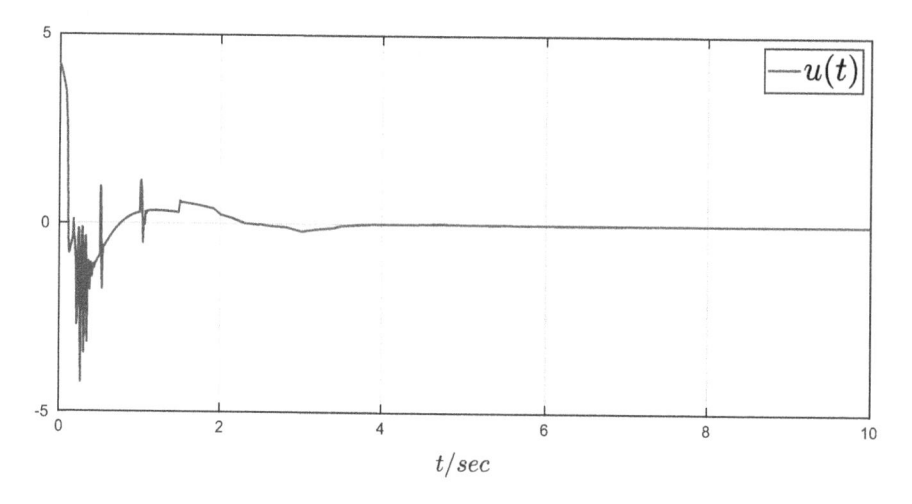

FIGURE 5.4 Control input $u(t)$.

performance measurement in the paper is considered for the sliding mode dynamics that matters the main interest. As we can see from these figures, by the adaptive SMC law (5.54), the estimated state $\hat{x}(t)$ converged to $x(t)$ quickly and achieved stochastic stability. Thus, the proposed method in this paper is effective.

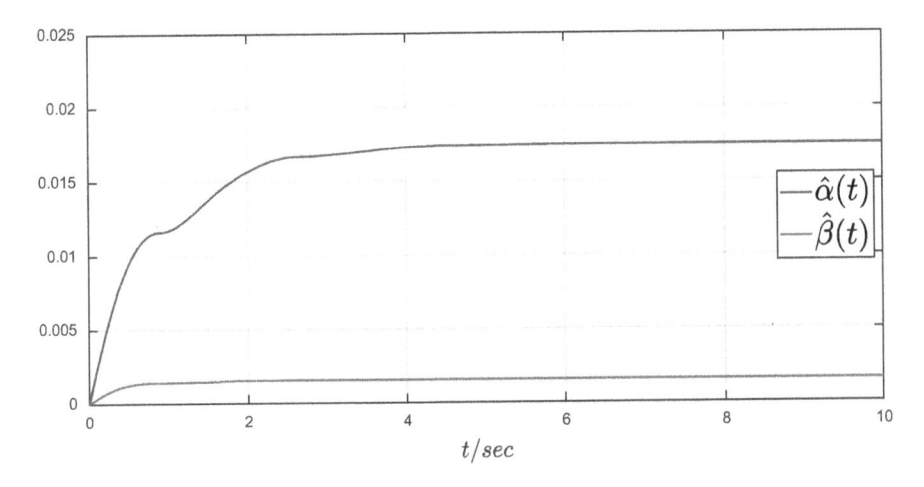

FIGURE 5.5 Estimated values of $\alpha(t)$ and $\beta(t)$.

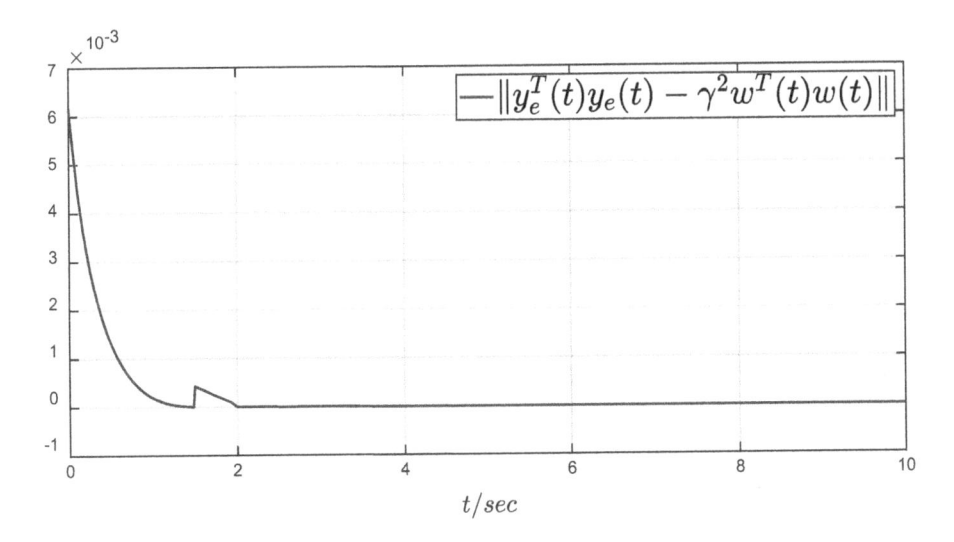

FIGURE 5.6 The integral item for the H_∞ measurement.

5.5 CONCLUSION

In this chapter, an observer-based adaptive ISMC strategy has been proposed for a class of nonlinear semi-Markovian jump T-S fuzzy systems with immeasurable premise variables. A novel integral sliding surface function has been put forward on observer space based on different input matrices. Feasible LMI conditions have been established to ensure the sliding mode dynamics and the error system with generally uncertain TRs to be stochastically stable with an H_∞ disturbance attenuation level γ.

Furthermore, an adaptive SMC law has been synthesized such that the sliding surface could be reached in finite time. Finally, a practical example has been provided to demonstrate the validity of the obtained results numerically.

REFERENCES

[1] D. W. C. Ho, and Y. Niu, Robust fuzzy design for nonlinear uncertain stochastic systems via sliding-mode control, *IEEE Transactions on Fuzzy Systems*, Vol. 15, No. 3, pp. 350–358, 2007.

[2] Q. Gao, G. Feng, Z. Xi, and Y. Wang, A new design of robust H_∞ sliding mode control for uncertain stochastic T-S fuzzy time-delay systems, *IEEE Transactions on Cybernetics*, Vol. 44, No. 9, pp. 1556–1566, 2014.

[3] Q. Gao, G. Feng, Z. Xi, Y. Wang, and J. Qiu, Robust H_∞ control of T-S fuzzy time-delay systems via a new sliding-mode scheme, *IEEE Transactions on Fuzzy Systems*, Vol. 22, No. 2, pp. 459–465, 2014.

[4] J. Li, Q. Zhang, X. G. Yan, and S. K. Spurgeon, Robust stabilization of T-S fuzzy stochastic descriptor systems via integral sliding modes, *IEEE Transactions on Cybernetics*, Vol. 48, No. 9, pp. 2736–2749, 2017.

[5] Y. Wang, Y. Gao, H. R. Karimi et al. Sliding mode control of fuzzy singularly perturbed systems with application to electric circuit, *IEEE Transactions on Systems, Man, and Cybernetics: Systems*, Vol. 48, No. 10, pp. 1667–1675, 2017.

[6] Y. Wang, H. Shen, H. R. Karimi, and D. P. Duan, Dissipativity-based fuzzy integral sliding mode control of continuous-time T-S fuzzy systems, *IEEE Transactions on Fuzzy Systems*, Vol. 26, No. 3, pp. 1164–1176, 2017.

[7] Z. Xi, G. Feng, and T. Hesketh, Piecewise integral sliding-mode control for T-S fuzzy systems, *IEEE Transactions on Fuzzy Systems*, Vol. 19, No.1, pp. 65–74, 2011.

[8] B. Jiang, H. R. Karimi, Y. Kao, C. Gao, A novel robust fuzzy integral sliding mode control for nonlinear semi-Markovian jump T-S fuzzy systems, *IEEE Transactions on Fuzzy Systems*, Vol. 26, No. 6, pp. 3594–3604, 2018.

[9] B. Jiang, H. R. Karimi, Y. Kao, C. Gao, Takagi-Sugeno model based sliding mode observer design for finite-time synthesis of semi-Markovian jump systems, *IEEE Transactions on Systems, Man, and Cybernetics: Systems*, Vol. 49, No. 7, pp. 1505–1515, 2018.

[10] H. Li, J. Wang, and P. Shi, Output-feedback based sliding mode control for fuzzy systems with actuator saturation, *IEEE Transactions on Fuzzy Systems*, Vol. 24, No. 6, pp. 1282–1293, 2016.

[11] F. Li, C. Du, C. Yang and W. Gui, Passivity-based asynchronous sliding mode control for delayed singular Markovian jump systems, *IEEE Transactions on Automatic Control*, Vol. 63, No. 8, pp. 2715–2721, 2017.

[12] B. Jiang, Y. Kao, C. Gao, and X. Yao, Passification of uncertain singular semi-Markovian jump systems with actuator failures via sliding mode approach, *IEEE Transactions on Automatic Control*, Vol. 62, No. 8, pp. 4138–4143, 2017.

[13 H. R. Karimi, A sliding mode approach to H_∞ synchronization of masterlave time-delay systems with Markovian jumping parameters and nonlinear uncertainties, *Journal of the Franklin Institute*, Vol. 349, No. 4, pp. 1480–1496, 2012.

[14] L. Wu, P. Shi, and H. Gao, State estimation and sliding-mode control of Markovian jump singular systems, *IEEE Transactions on Automatic Control*, Vol. 55, No. 5, pp. 1213–1219, 2010.

[15] Y. Kao, J. Xie, C. Wang, and H. R. Karimi, A sliding mode approach to H_∞ non-fragile observer-based control design for uncertain Markovian neutral-type stochastic systems, *Automatica*, Vol. 52, pp. 218–226, 2015.

[16] Q. Jia, W. Chen, Y. Zhang, and H. Li, Fault reconstruction and fault-tolerant control via learning observers in Takagi-Sugeno fuzzy descriptor systems with time delays, *IEEE Transactions on Industrial Electronics*, Vol. 62, No. 6, pp. 3885–3895, 2015.

[17] S. Huang, and G. Yang, Fault tolerant controller design for T-S fuzzy systems with time-varying delay and actuator faults: A k-step fault estimation approach, *IEEE Transactions on Fuzzy Systems*, Vol. 22, No. 6, pp. 1526–1540, 2014.

[18] M. Chadli, and H. R. Karimi, Robust observer design for unknown inputs Takagi-Sugeno models, *IEEE Transactions on Fuzzy Systems*, Vol. 21, No. 1, pp. 158–164, 2013.

[19] Z. Ning, L. Zhang and J. Lam, Stability and stabilization of a class of stochastic switching systems with lower bound of sojourn time, *Automatica*, Vol. 92, pp. 18–28, 2018.

[20] J. Song, Y. Niu and Y. Zou, Asynchronous sliding mode control of Markovian jump systems with time-varying delays and partly accessible mode detection probabilities, *Automatica*, Vol. 93, pp. 33–41, 2018.

[21] H. H. Choi, Robust stabilization of uncertain fuzzy systems using variable structure system approach, *IEEE Transactions on Fuzzy Systems*, Vol. 16, No. 3, pp. 715–724, 2008.

[22] H. N. Wu, and K. Y. Cai, Mode-independent robust stabilization for uncertain Markovian jump nonlinear systems via fuzzy control, *IEEE Transactions on Systems, Man, and Cybernetics: Systems*, Vol. 36, No. 3, pp. 509–519, 2005.

[23] R. Palm and D. Driankov, Fuzzy switched hybrid systems-modeling and identification, in *Proceedings of the 1998 IEEE. ISIC/CIRA/ISAS Joint Conference. Gaithersburg, MD*, 1998, pp. 130–135.

[24] K. Tanaka and T. Kosaki, Design of a stable fuzzy controller for an articulated vehicle, *IEEE Transactions on Systems, Man, and Cybernetics, Part B (Cybernetics)*, Vol. 27, No. 3, pp. 552–558, 1997.

6 Decentralized Adaptive Sliding Mode Control of Large-Scale Semi-Markovian Jump Systems

6.1 INTRODUCTION

Physical systems such as power networks, flexible communication networks and economics, etc. are often of high dimensions and complex structures; this kind of systems is the so-called large-scale systems (LSSs) [1,2]. Usually, LSSs are described by a set of interconnected low-order subsystems. Due to merits of the composition, decentralized control scheme [3] was then preferred to a centralized controller in the control synthesis of interconnected systems. Recently, we have witnessed a rapid growth in the application of decentralized control methodologies in engineering field: for instance, the linearized power system subject to different load profiles was investigated in Ref. [4]. A linear matrix inequality (LMI) approach was adopted to study decentralized control for multimachine power systems in Ref. [5] and robust decentralized control of power systems based only on swing angle measurements in Ref. [6], etc. However, it would be a challenge to design such a decentralized controller due to the difficulty that arises in handling these unknown interconnected terms from other subsystems. The decentralized sliding mode control (SMC) of LSSs also has been touched by some researchers in Refs. [7–9]. Unluckily, the interconnections among subsystems were assumed either exactly known or upper-bounded norm for the ease of controller design. What if the interconnected information is unknown? In fact, it is not an easy work to obtain this knowledge in practice because of the high complexity of system structure.

Based on the above discussion, this chapter concerns the issue of decentralized adaptive SMC scheme for the stabilization of large-scale semi-Markovian jump-interconnected systems, in which dead-zone characteristics are suffered in actuators. The difficulty lies in proposing a sliding mode controller that in the presence of nonlinear dead-zone inputs and cooperating SMC with adaptive laws such to ensure finite-time reachability of sliding surface and to remain a well formulated sliding motion. The novelties brought are that by designing local information-based sliding surface for each subsystem, the obtained sliding mode dynamics is also just local information-based with good property of dynamical performance. Last, the stabilization of each subsystem is realized through local controller, which facilitates the implementation process.

6.2 SYSTEM DESCRIPTION

Consider the following semi-Markovian jump-interconnected system $\mathcal{X}$ comprising N subsystems $\mathcal{X}_i$, $i \in \mathcal{N} = \{1,2...,N\}$. The dynamics of the i-th subsystem is described by:

$$\begin{cases} \dot{x}_i(t) = A_i(\eta_t)x_i(t) + B_i(\eta_t)(\phi_i(u_i(t)) + f_i(x_i,t)) + D_i(\eta_t)r_i(\mathbf{x}_j(t)) \\ x_i(0) = \varphi_{i0}, \eta_{t_0} = \eta_0, \end{cases} \tag{6.1}$$

where $x_i(t) \in \mathbb{R}^{n_i}$ is the state vector of the i-th subsystem, $x(t) = [x_1^T(t),...,x_N^T(t)]^T$ is the global state vector with $\sum_{i=1}^{N} n_i = n$; $\phi_i(u_i(t)$ is the nonlinear function with respect to control input $u_i(t) \in \mathbb{R}$; $x_i(0)$ is the vector-valued initial condition of the i-th subsystem. The matrices $A_i(\eta_t)$, $B_i(\eta_t)$ and $G_i(\eta_t)$ are of appropriate dimensions. The parameter uncertainty $f_i(x_i,t) \in \mathbb{R}$ satisfying $\| f_i(x_i,t) \| \leq \alpha_i \| x_i \|$ with α_i being a known positive constant. $r_i(\mathbf{x}_j(t))$ is the unknown interconnection input that describes the effect of other subsystems and satisfies

$$\| r_i(\mathbf{x}_j(t)) \| \leq \sum_{j \neq i}^{N} \beta_j \| x_j(t) \|,$$

in which $\mathbf{x}_j(t)$ is determined according to $r_i(\cdot)$ and defined by $\mathbf{x}_j(t) \triangleq [x_1^T(t) \cdots x_{i-1}^T(t) \ x_{i+1}^T(t) \cdots x_N^T(t)]^T$, which is the component of the state $x(t)$ that excludes i-th subsystem component $x_i(t)$, and β_j are unknown positive scalars. Generally, there exists a common $\beta_i > 0$ that can substitute all β_j for each subsystem.

$\{\eta_t, t \geq 0\}$ is a continuous-time semi-Markovian process taking discrete values in a finite set $S = \{1,2,...,s\}$ with generator given by

$$\Pr\{\eta_{t+h} = l \mid \eta_t = k\} = \begin{cases} \pi_{kl}(h)h + o(h), & k \neq l, \\ 1 + \pi_{kl}(h)h + o(h), & k = l, \end{cases} \tag{6.2}$$

where $h > 0$ and $\lim_{h \to 0} o(h)/h = 0$, $\pi_{kl}(h) > 0, k \neq l$, is the transition rate from mode k at time t to mode l at time $t+h$, and $\pi_{kk}(h) = -\sum_{l \neq k} \pi_{kl}(h) < 0$ for each $k \in S$.

Since $\pi_{kl}(h)$ is time-varying, it is not easy to acquire direct consideration of high complexity of LSSs. This part also considers the two cases: (I): $\pi_{kl}(h)$ is completely unknown; (II): $\pi_{kl}(h)$ is not exactly known but upper- and lower-bounded. For the case (II), it is further assumed that $\pi_{kl}(h) \in [\underline{\pi}_{kl}, \overline{\pi}_{kl}]$, in which $\underline{\pi}_{kl}$ and $\overline{\pi}_{kl}$ are the known real constants representing the lower and upper bounds of $\pi_{kl}(h)$, respectively. Moreover, denote $\pi_{kl}(h) \triangleq \pi_{kl} + \Delta\pi_{kl}(h)$, in which $\pi_{kl} = \frac{1}{2}(\underline{\pi}_{kl} + \overline{\pi}_{kl})$ and $|\Delta\pi_{kl}(h)| \leq \lambda_{kl}$ with $\lambda_{kl} = \frac{1}{2}(\overline{\pi}_{kl} - \underline{\pi}_{kl})$. So, the transition rate (TR) matrix with three jumping modes may be described as

$$\begin{bmatrix} \pi_{11} + \Delta\pi_{11}(h) & ? & \pi_{13} + \Delta\pi_{13}(h) \\ ? & ? & \pi_{23} + \Delta\pi_{23}(h) \\ ? & \pi_{32} + \Delta\pi_{32}(h) & ? \end{bmatrix}, \tag{6.3}$$

where "?" is the description of unknown TRs. For brevity, $\forall\, k \in \mathcal{S}$, let $I_k = I_{k,kn} \cup I_{k,ukn}$, where

$$I_{k,kn} \triangleq \{l : \pi_{kl} \text{ can be determined for } 1 \in \mathcal{S}\},$$

$$I_{k,ukn} \triangleq \{l : \pi_{kl} \text{ is not known for } l \in \mathcal{S}\}.$$

Similarly, we consider both $I_{k,kn} \neq \varnothing$ and $I_{k,ukn} \neq \varnothing$ and define the following set:

$$I_{k,kn} \triangleq \{\ell_{k,1}, \ell_{k,2}, \dots, \ell_{k,o}\} \quad 1 \leq o < s,$$

in which o denotes the number of elements in the set, and $\ell_{k,\iota}$ ($\iota = 1,2,\dots,o\}$) represents the index of i-th element in the k-th row of the TR matrix.

Remark 6.1

Note that in Ref. [1], different mode process $\eta_i(t)$ is adopted for each i-th subsystem. But there is a bijective mapping $\psi : \mathcal{M}_v \rightarrow \mathcal{M}_s$ that transforms these individual processes $\eta_i(t)$ into a global jumping process $\eta(t)$. $\mathcal{M}_v$ is a set in which $[\eta_1(t) \dots \eta_N(t)]$ takes values, where $\eta_i(t) \in \{1,2,\dots,M_i\} i \in \mathcal{N}$. $\mathcal{M}_s$ is a set defined by $\mathcal{M}_s \triangleq \{1,2,\dots,M\}$ with $M \leq \prod_{i=1}^{N} M_i$.

For the nonlinear LSS (6.1), the stabilization purpose is realized by implementing the actuating device $\phi_i(u_i(t))$, which suffers from the dead-zone characteristics. $u_i(t)$ is the control input which will be designed later. The subsystem diagram is shown in Figure 6.1. In order to clarify the characteristics of the dead-zone nonlinearity $\phi_i(u_i(t))$, it is defined in the following.

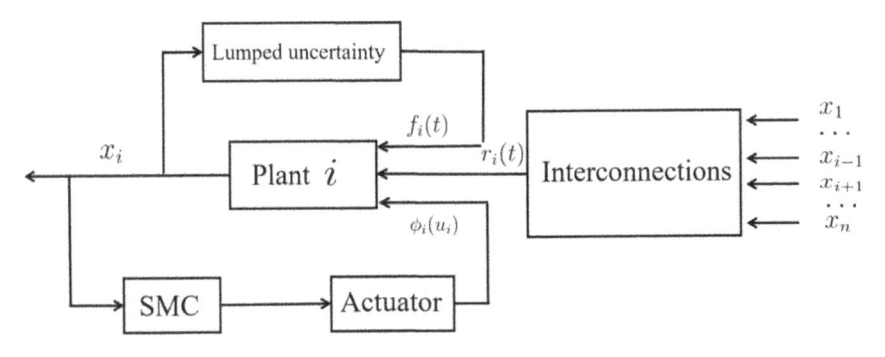

FIGURE 6.1 The diagram of subsystem $\mathcal{X}_i$.

Definition 6.1 [10]

The allowable dead-zone nonlinear input function $\phi_i(u_i(t))$ satisfies that

$$\phi_i(u_i) = \begin{cases} \bar{\phi}_i(u_i)(u_i - \bar{u}_i) & u_i > \bar{u}_i \\ 0 & -\underline{u}_j \leq u_i \leq \bar{u}_i, \\ \underline{\phi}_i(u_i)(u_i + \underline{u}_j) & u_i < -\underline{u}_j \end{cases} \tag{6.4}$$

where $\bar{\phi}_i(u_i) > 0$ and $\underline{\phi}_i(u_i) > 0$ are nonlinear functions of u_i. $\bar{u}_i > 0$ and $\underline{u}_j > 0$ are constants.

Therefore, $\phi_i(u_i)$ satisfies the following properties:

$$(u_i - \bar{u}_i)\phi_i(u_i) \geq \bar{m}_i(u_i - \bar{u}_i)^2, \quad u_i > \bar{u}_i$$

$$(u_i + \underline{u}_j)\phi_i(u_i) \geq \underline{m}_i(u_i + \underline{u}_j)^2, \quad u_i < -\underline{u}_j \tag{6.5}$$

where $\bar{m}_i \leq \bar{\phi}_i(u_i)$ and $\underline{m}_i \leq \underline{\phi}_i(u_i)$ are the constants. Without the loss of generality, in this paper, we consider the case $\phi_i(u_i)$ is antisymmetric about the origin, which follows $\bar{u}_i = \underline{u}_j = u_{i0}$ and $\bar{m}_i = \underline{m}_i = m_i$. Based on this condition, (6.5) is rewritten as

$$(u_i - u_{i0})\phi_i(u_i) \geq \bar{m}_i(u_i - u_{i0})^2, \quad u_i > u_{i0}$$

$$(u_i + u_{i0})\phi_i(u_i) \geq \underline{m}_i(u_i + u_{i0})^2, \quad u_i < -u_{i0} \tag{6.6}$$

Assumption 6.1

The input matrix $B_i(k)$ of each subsystem is of full column rank, and there exists matrix $H_i(k)$ with appropriate dimensions such that $D_i(k) = B_i(k)H_i(k)$.

Definition 6.2 [1]

The semi-Markovian jump-interconnected system with N subsystems is stochastically stable if it holds

$$\mathbf{E}\left\{\int_0^\infty \left(\sum_{i=1}^N \|x_i(t)\|^2\right)dt \mid x_0, \eta_0\right\} < \infty,$$

where initial condition $x(0) = [x_1^T(0),\ldots,x_N^T(0)]^T$, $\eta_0 \in \mathcal{S}$ and $i \in \mathcal{N}$.

The main objective of this chapter is, without the requirement of the information of the unknown interconnections and nonlinear perturbations, to design a model-free reference adaptive sliding mode controller such that the states of each subsystem could be driven onto sliding surface and maintain sliding motion. In addition, the secondary objective is to guarantee the stochastic stability of the interconnected LSS (6.1) with generally uncertain TRs.

6.3 MAIN RESULTS

For the LSS (6.1), two steps are involved in the design procedure of sliding mode controller. First, for subsystems, it is needed to design suitable sliding surfaces, on which the sliding motion happens and the controlled system has the desired dynamic performance. The composite switching surface of the proposed control scheme is defined by letting the composite vector S to be zero, that is, $S = [s_1^T \ s_2^T \ \cdots \ s_N^T]^T = 0$.

6.3.1 SLIDING SURFACE DESIGN

Let the associated sliding surface for each subsystem be defined as

$$s_i(t) = G_i x_i(t) - \int_0^t G_i(A_i(k) + B_i(k)K_i(k))x_i(s)ds, \qquad (6.7)$$

where $G_i \in \mathbb{R}^{n_i}$ is chosen such that $G_i B_i(k)$ is nonsingular, and $K_i(k) \in \mathbb{R}^{n_i}$ is adopted such that $A_i(k) + B_i(k)K_i(k)$ is Hurwitz.

According to variable structure control theory [11], in the sliding hyperplane, there are $s_i(t) = 0$ and $\dot{s}_i(t) = 0$, respectively. By the condition $\dot{s}_i(t) = 0$, that is,

$$\dot{s}_i(t) = G_i[B_i(k)(\phi_i(u_i(t)) + f_i(x_i,t))$$
$$+ D_i(k)r_i(\mathbf{x}_j(t)) - B_i(k)K_i(k)x_i(t)] = 0 \qquad (6.8)$$

Therefore, it obtains the following equivalent control from (6.8),

$$\phi_i(u_{eq}) = K_i(k)x_i(t) - f_i(x_i,t) - (G_i B_i(k))^{-1} G_i D_i(k)r_i(\mathbf{x}_j(t)). \qquad (6.9)$$

Substituting (6.9) into (6.1) and applying the condition in Assumption 6.1, it derives the following sliding mode dynamics:

$$\dot{x}_i(t) = (A_i(k) + B_i(k)K_i(k))x_i(t). \qquad (6.10)$$

After deriving the sliding mode dynamics (6.10), the next phase is to analyze the stochastic stability in the sense of Definition 6.2.

6.3.2 STABILITY ANALYSIS OF SLIDING MODE DYNAMICS

Since the memoryless property in the MJSs does not pertain in S-MJSs, the TR has the properties of time-varying or high nonlinear, which poses the main technical difficulty in proposing stochastic stability criteria when taking into account generally uncertain TRs. However, the following theorem provides stochastic stability analysis of LSS with semi-Markovian jump parameters based on Ref. [12].

Theorem 6.1

The dynamics (6.1) with subsystems in possession of sliding mode dynamics (6.10) is stochastically stable if there exist positive-definite matrices $P_i(k) > 0$, $U_i(kl) > 0$ and $V_i(kl) > 0$ such that the following conditions hold for $k \in \mathcal{S}, i \in \mathcal{N}$

If $k \in I_{k,kn}$, $\forall j \in I_{k,ukn}$, $I_{k,kn} \triangleq \{\ell_{k,1}, \ell_{k,2}, \dots, \ell_{k,o_1}\}$,

$$
\begin{bmatrix}
\mathcal{A}_i^{11}(k) & \mathcal{A}_i^{12}(k) \\
* & \mathcal{A}_i^{22}(k)
\end{bmatrix} < 0,
\tag{6.11}
$$

If $k \in I_{k,ukn}$, $\forall j \in I_{k,ukn}$, $I_{k,kn} \triangleq \{\ell_{k,1}, \ell_{k,2}, \dots, \ell_{k,o_2}\}, j \neq k$,

$$
P_i(k) - P_i(j) \geq 0
\tag{6.12a}
$$

$$
\begin{bmatrix}
\mathcal{A}_i^{21}(k) & \mathcal{A}_i^{22}(k) \\
* & \mathcal{A}_i^{23}(k)
\end{bmatrix} < 0,
\tag{6.12b}
$$

where

$$
\mathcal{A}_i^{11}(k) = \mathrm{He}\{P_i(k)(A_i(k) + B_i(k)K_i(k))\} + \sum_{l \in I_{k,kn}} \left[\frac{(\lambda_{kl})^2}{4} U_i(k,l) + \pi_{kl}(P_i(l) - P_i(j)) \right],
$$

$$
\mathcal{A}_i^{12}(k) = [(P_i(\ell_{k,1}) - P_i(j)) \ \dots \ (P_i(\ell_{k,o_1}) - P_i(j))],
$$

$$
\mathcal{A}_i^{13}(k) = [-U_i(k,\ell_{k,1}) \ \dots \ -U_i(k,\ell_{k,o_1})],
$$

$$
\mathcal{A}_i^{21}(k) = \mathrm{He}\{P_i(k)(A_i(k) + B_i(k)K_i(k))\} + \sum_{l \in I_{k,kn}} \left[\frac{(\lambda_{kl})^2}{4} V_i(k,l) + \pi_{kl}(P_i(l) - P_i(j)) \right],
$$

$$
\mathcal{A}_i^{22}(k) = [(P_i(\ell_{k,1}) - P_i(j)) \ \dots \ (P_i(\ell_{k,o_2}) - P_i(j))],
$$

$$
\mathcal{A}_i^{23}(k) = [-V_i(k,\ell_{k,1}) \ \dots \ -V_i(k,\ell_{k,o_2})].
$$

Proof: Consider the Lyapunov functional of the form:

$$
V(t) = \sum_{i=1}^{N} V_i(x_i(t), \eta_t) = \sum_{i=1}^{N} x_i^T(t) P_i(\eta_t) x_i(t).
\tag{6.13}
$$

Then, according to Definition 1.5 and the method proposed in Ref. [12], we have

$$\mathcal{L}V_i(x_i(t),k) = \lim_{\delta \to 0} \frac{1}{\delta} \left[\sum_{l=1,l\neq k}^{s} \Pr\{\eta_{t+\delta} = l \mid \eta_t = k\} x_{i,\delta}^T P_i(l) x_{i,\delta} \right.$$

$$\left. + \Pr\{\eta_{t+\delta} = l \mid \eta_t = k\} x_{i,\delta}^T P_i(k) x_{i,\delta} - x_i^T(t) P_i(k) x_i(t) \right], \qquad (6.14)$$

where $x_{i,\delta} \triangleq x_i(t+\delta)$. For a general distribution of the sojourn time without memoryless property, that is, $\Pr\{\eta_{t+\delta} = l \mid \eta_t = k\} \neq \Pr\{\eta_\delta = l \mid \eta_0 = k\}$, by the conditional probability formula, we have

$$\mathcal{L}V_i(x_i(t),k) = \lim_{\delta \to 0} \frac{1}{\delta} \left[\sum_{l=1,l\neq k}^{s} \frac{q_{kl}(G_k(h+\delta)-G_k(t))}{1-G_k(h)} x_{i,\delta}^T P_i(l) x_{i,\delta} \right.$$

$$\left. + \frac{1-G_k(h+\delta)}{1-G_k(h)} x_{i,\delta}^T P_i(k) x_{i,\delta} - x_i^T(t) P_k x_i(t) \right]$$

$$= \lim_{\delta \to 0} \frac{1}{\delta} \left[\sum_{l=1,l\neq k}^{s} \frac{q_{kl}(G_k(h+\delta)-G_k(h))}{1-G_k(h)} x_{i,\delta}^T P_i(l) x_{i,\delta} \right.$$

$$+ \frac{1-G_k(h+\delta)}{1-G_k(h)} [x_{i,\delta}^T - x_i^T(t)] P_i(k) x_{i,\delta}$$

$$+ \frac{1-G_k(h+\delta)}{1-G_k(h)} x_i^T(t) P_i^T(k) [x_{i,\delta} - x_i(t)]$$

$$\left. - \frac{G_k(h+\delta)-G_k(h)}{1-G_k(h)} x_i^T(t) P_i(k) x_i(t) \right] \qquad (6.15)$$

where $G_k(h)$ is defined as the cumulative distribution function of the sojourn time when the system stays in mode k and q_{kl} is the probability intensity from mode k to mode l. On the other hand, we have that

$$\lim_{\delta \to 0} \frac{(G_k(h+\delta)-G_k(h))}{(1-G_k(h))\delta} = \pi_k(h),$$

$$\lim_{\delta \to 0} \frac{1-G_k(h+\delta)}{1-G_k(h)} = 1, \qquad (6.16)$$

where $\pi_k(h)$ is the TR of the system jumping from mode k. Therefore,

$$\mathcal{L}V_i(x_i(t),k) = \sum_{l=1,l\neq k}^{s} q_{kl}\pi_k(h) x_i^T(t) P_i(l) x_i(t)$$

$$+ 2x_i^T(t) P_i(k) \dot{x}(t) - \pi_k(h) x_i^T(t) P_i(k) x_i(t). \qquad (6.17)$$

Now, define $\pi_{kl}(h) = \pi_k(h)q_{kl}$ for $l \neq k$ and $\pi_{kk}(h) = -\sum_{l \neq k} \pi_{kl}(h)$. Overall, we have

$$\mathcal{L}V(t) = \mathcal{L}\sum_{i=1}^{N} V_i(x_i(t), k) = \sum_{i=1}^{N} x_i^T(t)\check{\Gamma}_i(k)x_i(t), \qquad (6.18)$$

where $\check{\Gamma}_i(k) = \Gamma_i(k) + \sum_{l=1}^{s} \pi_{kl}(h)P_i(l)$ with $\Gamma_i(k) = He\{P_i(k)(A_i(k) + B_i(k)K_i(k))\}$.

Therefore, if $\check{\Gamma}_i(k) < 0$, then the globally interconnected semi-Markovian jump system is stochastically stable. So, the following proof continues. Let us consider the following two cases:

Case I: $k \in I_{k,kn}$.

First, denote $\lambda_{k,kn} \triangleq \sum_{l \in I_{k,kn}} \pi_{kl}(h)$. Since $I_{k,ukn} \neq \varnothing$, it holds that $\lambda_{k,kn} < 0$. Notice that $\sum_{l=1}^{s} \pi_{kl}(h)P_i(l)$ can be represented as

$$\sum_{l=1}^{s} \pi_{kl}(h)P_i(l) = \left(\sum_{l \in I_{k,kn}} + \sum_{l \in I_{k,ukn}}\right)\pi_{kl}(h)P_i(l)$$

$$= \sum_{l \in I_{k,kn}} \pi_{kl}(h)P_i(l) - \lambda_{k,kn} \sum_{l \in I_{k,ukn}} \frac{\pi_{kl}(h)}{-\lambda_{k,kn}}P_i(l), \qquad (6.19)$$

and it is obvious that $0 \leq \pi_{kl}(h)/-\lambda_{k,kn} \leq 1$ $(l \in I_{k,ukn})$ and $\sum_{l \in I_{k,ukn}} \frac{\pi_{kl}(h)}{-\lambda_{k,kn}} = 1$. So for $\forall j \in I_{k,ukn}$, there is

$$\check{\Gamma}_i(k) = \sum_{l \in I_{k,ukn}} \frac{\pi_{kl}(h)}{-\lambda_{k,kn}}\left[\Gamma_i(k) + \sum_{l \in I_{k,kn}} \pi_{kl}(h)(P_i(l) - P_i(j))\right]. \qquad (6.20)$$

Therefore, for $0 \leq \pi_{kl}(h) \leq -\lambda_{k,kn}, \check{\Gamma}_i(k) < 0$ is equivalent to

$$\Gamma_i(k) + \sum_{l \in I_{k,kn}} \pi_{kl}(h)(P_i(l) - P_i(j)) < 0. \qquad (6.21)$$

In the formula (6.21), it is true that

$$\sum_{l \in I_{k,kn}} \pi_{kl}(h)(P_i(l) - P_i(j)) = \sum_{l \in I_{k,kn}} \pi_{kl}(P_i(l) - P_i(j))$$

$$+ \sum_{l \in I_{k,kn}} \Delta\pi_{kl}(h)(P_i(l) - P_i(j)). \qquad (6.22)$$

Then, by virtue of Lemma 1.4 and for any $U_i(kl) > 0$, it follows that

$$\sum_{l \in I_{k,kn}} \Delta \pi_{kl}(h)(P_i(l) - P_i(j))$$

$$= \sum_{l \in I_{k,kn}} \left[\frac{1}{2} \Delta \pi_{kl}(h)((P_i(l) - P_i(j)) + (P_i(l) - P_i(j))) \right]$$

$$\leq \sum_{l \in I_{k,kn}} \left[\frac{(\lambda_{kl})^2}{4} U_i(kl) + (P_i(l) - P_i(j))U_i(kl))^{-1}(P_i(l) - P_i(j))^T \right]. \qquad (6.23)$$

Combining (6.19)–(6.23), then by applying Schur complement, it is deduced that (6.11) guarantees $\check{\Gamma}_{l,m} < 0$ when $k \in I_{k,kn}$.

Case II: $k \in I_{k,ukn}$.

Similarly, denote $\lambda_{k,kn} \triangleq \sum_{l \in I_{k,kn}} \pi_{kl}(h)$. Since $I_{k,kn} \neq \varnothing$, it holds that $\lambda_{k,kn} > 0$. Now $\sum_{l=1}^{s} \pi_{kl}(h)P_i$ can be represented as

$$\sum_{l=1}^{s} \pi_{kl}(h)P_i(l) = \sum_{l \in I_{k,kn}} \pi_{kl}(h)P_i(l) + \pi_{kk}(h)P_i(k) + \sum_{l \in I_{k,ukn}, l \neq k} \pi_{kl}(h)P_i(l)$$

$$= \sum_{l \in I_{k,kn}} \pi_{kl}(h)P_i(l) + \pi_{kk}(h)P_i(k)$$

$$- (\pi_{kk}(h) + \lambda_{k,kn}) \sum_{l \in I_{k,ukn}, l \neq k} \frac{\pi_{kl}(h)P_i(l)}{-\pi_{kk}(h) - \lambda_{k,kn}}, \qquad (6.24)$$

and it is obvious that $0 \leq \pi_{kl}(h)/-\pi_{kk}(h) - \lambda_{k,kn} \leq 1$ ($l \in I_{k,ukn}$) and $\sum_{l \in I_{k,ukn}, l \neq k} \frac{\pi_{kl}(h)}{-\pi_{kk}(h) - \lambda_{k,kn}} = 1$. So for $\forall j \in I_{k,ukn}, j \neq k$,

$$\check{\Gamma}_i(k) = \sum_{l \in I_{k,ukn}, l \neq k} \frac{\pi_{kl}(h)}{-\pi_{kk}(h) - \lambda_{k,kn}} \left[\Gamma_i(k) + \pi_{kk}(h)(P_i(k) - P_i(j)) \right.$$

$$\left. + \sum_{l \in I_{k,kn}} \pi_{kl}(h)(P_i(l) - P_i(j)) \right]. \qquad (6.25)$$

Therefore, for $0 \leq \pi_{kl}(h) \leq -\pi_{kk}(h) - \lambda_{k,kn}$, $\check{\Gamma}_i(k) < 0$ is equivalent to

$$\Gamma_i(k) + \pi_{kk}(h)(P_i(k) - P_i(j)) + \sum_{l \in I_{k,kn}} \pi_{kl}(h)(P_i(l) - P_i(j)) < 0. \qquad (6.26)$$

Since $\pi_{kk}(h) < 0$, (6.26) holds if we have

$$\begin{cases} P_i(k) - P_i(j) \geq 0, \\ \Gamma_i(k) + \displaystyle\sum_{l \in I_{k,kn}} \pi_{kl}(h)(P_i(l) - P_i(j)) < 0. \end{cases} \tag{6.27}$$

Also, as in (6.22) and (6.23), for any $V_i(kl) > 0$, we have

$$\sum_{l \in I_{k,kn}} \pi_{kl}(h)(P_i(l) - P_i(j)) \leq \sum_{l \in I_{k,kn}} \pi_{kl}(P_i(l) - P_i(j)) + \sum_{l \in I_{k,kn}} \left[\frac{(\lambda_{kl})^2}{4} V_i(kl) \right.$$

$$\left. + (P_i(l) - P_i(j))V_i^{-1}(kl)(P_i(l) - P_i(j))^T \right]. \tag{6.28}$$

Combining (6.24)–(6.28), we know that (6.12a) and (6.12b) guarantee $\dot{\Gamma}_i(k) < 0$ by applying Schur complement when $k \in I_{k,ukn}$. In summary, the globally interconnected large-scale semi-Markovian jump system is stochastically stable. This completes the proof.

Remark 6.2

Although the TRs are time-varying and not exactly known, strict LMI conditions are established in Theorem 6.1 thanks to the average method applied in modeling the TR matrix, in which the constant ones are used directly and the uncertainties are being tackled in (6.23) and (6.28) by virtue of Lemma 1.4. Therefore, the overall criteria are combined into a unit instead of applying upper and lower bounds of uncertain TRs separately.

6.3.3   CONVERGENCE OF SLIDING SURFACE

In the above sections, we have designed the sliding surface and analyzed the stochastic stability of the corresponding sliding mode dynamics. The next step is to design an appropriate sliding mode controller such that the state of the controlled system will be attracted onto the predesigned sliding surface, and remain them there for all subsequent time.

It is well known by introducing SMC, the sliding mode is disturbance rejection and insensitive to plant parameter variations. However, the interconnections between subsystems are not known. Therefore, we employ an adaptive gain $\hat{\beta}_i(t)$ to adapt β_i in these unknown interconnections. The corresponding adaption error is defined by $\tilde{\beta}_i(t) = \hat{\beta}_i(t) - \beta_i$.

Theorem 6.2

Suppose that the sliding surface function is proposed in (6.7), and the conditions in Theorem 6.1 are feasible. Then, the state trajectories of the controlled system (6.1) will be almost surely driven onto the sliding surface $s(t) = 0$ in finite time by the adaptive SMC law synthesized as follows:

$$u_i(t) = -\frac{s_i(t)}{\| s_i(t) \|}(\rho_i(t) + u_{i0}), \tag{6.29}$$

where

$$\rho_i(t) = \frac{1}{m_i}[(\| K_i(k) \| + \alpha_i) \| x_i(t) \| + \sum_{j \neq i} \hat{\beta}_i(t) \| H_j(k) \| \| x_i(t) \| + m_i \sigma_i]$$

with $\sigma_i > 0$ being a small tuning scalar, and the adaptive gain is designed by

$$\dot{\hat{\beta}}_i(t) = c_{i0} \sum_{j \neq i} \| H_j(k) \| \| x_i(t) \|,$$

where c_{i0} is a positive scalar specified by the designer.

Proof: Choose the following Lyapunov function:

$$V(t) = \sum_{i=1}^{N}(s_i^T(t)s_i(t))^{1/2} + \sum_{i=1}^{N}\frac{1}{2}c_{i0}^{-1}\tilde{\beta}_i^2(t). \tag{6.30}$$

Then, we can derive the following equality:

$$\mathcal{L}V(t) = \sum_{i=1}^{N}\frac{[\dot{s}_i^T(t)s_i(t) + s_i^T(t)\dot{s}_i(t)]}{2\sqrt{s_i^T(t)s_i(t)}} + \sum_{i=1}^{N}c_{i0}^{-1}\tilde{\beta}_i(t)\dot{\tilde{\beta}}_i(t)$$

$$= \sum_{i=1}^{N}\frac{s_i(t)}{\| s_i(t) \|}G_iB_i(k)[(\phi_i(u_i(t)) + f_i(x_i, t))$$

$$+ H_i(k)r_i(x_j(t)) - K_i(k)x_i(t)] + \sum_{i=1}^{N}c_{i0}^{-1}\dot{\hat{\beta}}_i(t)(\hat{\beta}_i(t) - \beta_i). \tag{6.31}$$

Note that $\dot{\tilde{\beta}}_i(t) = \dot{\hat{\beta}}_i(t)$, and

$$\sum_{i=1}^{N}\frac{s_i(t)}{\| s_i(t) \|}G_iB_i(k)H_i(k)\sum_{j=1,j\neq i}^{N}\beta_i \| x_j(t) \|$$

$$= \sum_{i=1}^{N}\sum_{j=1,j\neq i}^{N}\frac{s_j(t)}{\| s_j(t) \|}G_jB_j(k)H_j(k)\beta_j \| x_i(t) \|. \tag{6.32}$$

Particularly, for simplicity, it is assumed there exist G_i such that $G_iB_i(k) = 1$. Therefore, based on the above equalities, the following inequality is derived from (6.31)

$$\mathcal{L}V(t) \leq \sum_{i=1}^{N} \frac{s_i(t)}{\| s_i(t) \|} \Big[\phi_i(u_i(t)) + \alpha_i \| x_i(t) \|$$

$$+ \left(\sum_{j=1, j \neq i}^{N} \beta_j \| H_j(k) \| + \| K_i(k) \| \right) \| x_i(t) \| \Big] + \sum_{i=1}^{N} c_{i0}^{-1} \dot{\hat{\beta}}_i(t)(\hat{\beta}_i(t) - \beta_i). \quad (6.33)$$

Changing the form of the control (6.29), that is,

$$u_i(t) + \frac{s_i(t)}{\| s_i(t) \|} u_{i0} = -\frac{s_i(t)}{\| s_i(t) \|} \rho_i(t). \quad (6.34)$$

Substituting the above equality into (6.6), we have

$$\frac{s_i(t)}{\| s_i(t) \|} \phi_i(u_i) \leq -m_i \rho_i(t). \quad (6.35)$$

Combining (6.33) and (6.35), and taking $\rho_i(t)$ and $\dot{\hat{\beta}}_i(t)$ in (6.29) into consideration, it overall leads to

$$\mathcal{L}V(t) \leq -\sum_{i=1}^{N} \sigma_i = -\sigma, \quad for \quad \| s_i(t) \| \neq 0. \quad (6.36)$$

Now, by integrating (6.36) from 0 to t and then taking expectation for both sides, one can get

$$\sum_{i=1}^{N} \mathbf{E} \| s_i(t) \| \leq \mathbf{E}V(t) \leq \mathbf{E}V(0) - \sigma t. \quad (6.37)$$

From (6.37), it derives $\mathbf{E} \| s_i(t) \| = 0$ for all $t \geq t^* = \mathbf{E}V(0)/\sigma$, which implies $s_i(t) = 0$ is almost surely guaranteed for $t \geq t^*$. This completes the proof.

Remark 6.3

Here, the unknown interconnections among subsystems are compensated by the designed adaptive law based on its local information, which is one of the main contributions comparing with existing literatures. Therefore, the stabilization of the global LSS is also realized by implementing local controllers, which alleviates the complexity in practice. In addition, the adaptive SMC successfully guarantees finite-time reachability of sliding surface and keeps a sliding motion along the dynamics (6.10).

6.4 NUMERICAL EXAMPLE

In this part, a numerical example is used to show the merits of the proposed results. Consider a semi-Markovian jump-interconnected system with two subsystems. Let $\eta(t)$ be the mode information of each subsystem that has two modes. The system data of each subsystem are given by

$$A_1(1) = \begin{bmatrix} -1 & 0.5 \\ -0.5 & 0 \end{bmatrix}, A_1(2) = \begin{bmatrix} 0.5 & -1 \\ 1 & 0.5 \end{bmatrix},$$

$$A_1(3) = \begin{bmatrix} 1 & -0.2 \\ 0 & 1 \end{bmatrix}, A_2(1) = \begin{bmatrix} 1 & -0.2 \\ 0.5 & -1.5 \end{bmatrix},$$

$$A_2(2) = \begin{bmatrix} 0.8 & -1 \\ 2 & 0.5 \end{bmatrix}, A_2(3) = \begin{bmatrix} 0.5 & 0 \\ -0.1 & -1 \end{bmatrix},$$

$$B_1(k) = \begin{bmatrix} 0 \\ 2 \end{bmatrix}, B_2(k) = \begin{bmatrix} 1 \\ 0 \end{bmatrix}, k = 1,2,3,$$

$$D_1(1) = \begin{bmatrix} 0 & 0 \\ 1 & 2 \end{bmatrix}, D_1(2) = \begin{bmatrix} 0 & 0 \\ 4 & 2 \end{bmatrix},$$

$$D_1(3) = \begin{bmatrix} 0 & 0 \\ 3 & 2 \end{bmatrix}, D_2(1) = \begin{bmatrix} 2 & 3 \\ 0 & 0 \end{bmatrix},$$

$$D_2(2) = \begin{bmatrix} 2 & 4 \\ 0 & 0 \end{bmatrix}, D_2(3) = \begin{bmatrix} 1 & 2 \\ 0 & 0 \end{bmatrix}.$$

The dead-zone nonlinear input functions are given by

$$\phi_1(u_1) = \begin{cases} (1.2 - 0.5e^{0.3|\sin u_1(t)|})(u_1 - u_{10}), & u_1 > u_{10} \\ 0, & |u_1| \le u_{10} \\ (1.2 - 0.5e^{0.3|\sin u_1(t)|})(u_1 + u_{10}), & u_1 < -u_{10} \end{cases},$$

$$\phi_2(u_2) = \begin{cases} (1.5 - 0.3e^{0.3|\cos u_2(t)|})(u_2 - u_{20}), & u_2 > u_{20} \\ 0, & |u_2| \le u_{20} \\ (1.5 - 0.3e^{0.3|\cos u_2(t)|})(u_2 + u_{20}), & u_2 < -u_{20} \end{cases}.$$

The corresponding TR matrix is described by

$$
\begin{bmatrix}
-2.0 + \Delta\pi_{11}(h) & \nabla_{12} & \nabla_{13} \\
\nabla_{21} & \nabla_{22} & 0.5 + \Delta\pi_{23}(h) \\
0.5 + \Delta\pi_{31}(h) & \nabla_{32} & -1.5 + \Delta\pi_{33}(h)
\end{bmatrix}.
$$

If the stochastic jumping parameters for each subsystem are different, we can follow a similar way employed in Ref. [1], and denote the augmented vector $[\eta_1(t), \eta_2(t)]^T$ as the global mode information of the interconnected system which takes values in $\mathcal{M}_v = \{[1,1]^T, [1,2]^T, [2,1]^T, [2,2]^T\}$. Then, through the bijective mapping ψ, we can define a global jump process $\eta(t)$ by taking values in a set $\mathcal{M}_s = \{1,2,3,4\}$.

For this work, the following parameters are given. $\alpha_1 = 0.5$, $\alpha_2 = 0.8$, $m_1 = m_2 = 0.5$, $u_{10} = u_{20} = 2$, $G_1 = [0 \ 0.5]$, $G_2 = [1 \ 0]$, $K_1(1) = K_1(2) = K_1(3) = [1 \ -1]$ and $K_2(1) = K_2(2) = K_2(3) = [-5 \ 0]$ are taken for simulation study. First, let's check the conditions in Theorem 6.1; it obtains the following feasible solutions:

$$
P_1(1) = \begin{bmatrix} 2.9396 & -0.1178 \\ -0.1178 & 0.8790 \end{bmatrix}, \ P_1(2) = \begin{bmatrix} 4.5799 & -0.9887 \\ -0.9887 & 1.6206 \end{bmatrix},
$$

$$
P_1(3) = \begin{bmatrix} 2.9719 & -0.3194 \\ -0.3194 & 1.2816 \end{bmatrix}, \ P_2(1) = \begin{bmatrix} 0.6433 & 0.5188 \\ 0.5188 & 1.3794 \end{bmatrix},
$$

$$
P_2(2) = \begin{bmatrix} 1.6110 & 1.0967 \\ 1.0967 & 1.8549 \end{bmatrix}, \ P_2(3) = \begin{bmatrix} 0.7261 & 0.3987 \\ 0.3987 & 1.0326 \end{bmatrix}.
$$

$$
U_1(31) = \begin{bmatrix} 3.2294 & -0.2236 \\ -0.2236 & 3.0299 \end{bmatrix}, \ U_1(33) = \begin{bmatrix} 3.1961 & -0.1439 \\ -0.1439 & 2.9215 \end{bmatrix},
$$

$$
U_2(11) = \begin{bmatrix} 3.0076 & 0.0881 \\ 0.0881 & 2.9261 \end{bmatrix}, \ U_2(31) = \begin{bmatrix} 3.1172 & 0.1984 \\ 0.1984 & 3.0392 \end{bmatrix},
$$

$$
U_2(33) = \begin{bmatrix} 3.1957 & 0.3873 \\ 0.3873 & 3.2922 \end{bmatrix}.
$$

For simulation purposes, the initial conditions and external disturbances are given as $x_1(0) = x_2(0) = [1 \ -1]^T$, $f_1(x_1, t) = 0.5 \sin x_{12}(t)$, $f_2(x_2, t) = 0.5 \cos x_{22}(t)$ and $\sigma_i = 0.01$. The adaptive gains are chosen as $c_{10} = c_{20} = 0.1$, and the interconnection terms are selected as $r_1(\mathbf{x}_2) = x_1^2$ and $r_2(\mathbf{x}_1) = x_2^2$. With the above parameters, Figure 6.2 shows that the open-loop system is unstable. After employing the SMC strategy, the closed-loop system achieves the prescribed stability property. Simulation

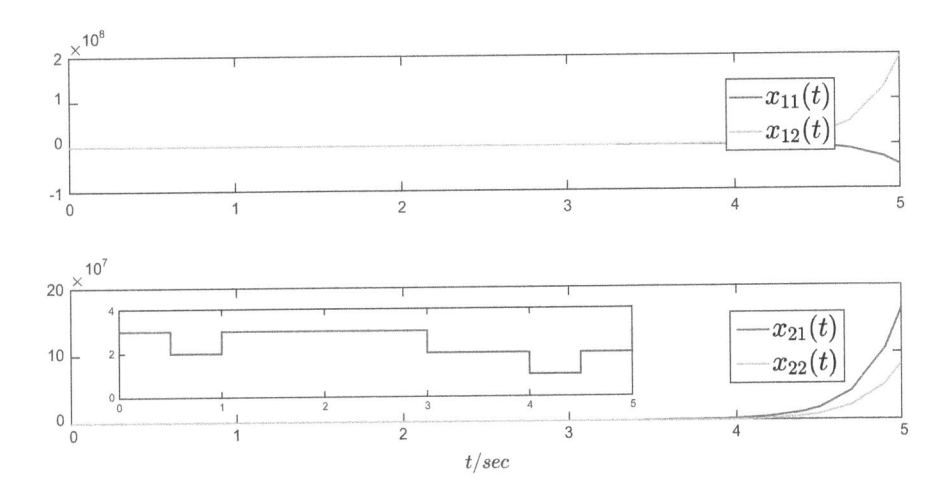

FIGURE 6.2 State response of open-loop system.

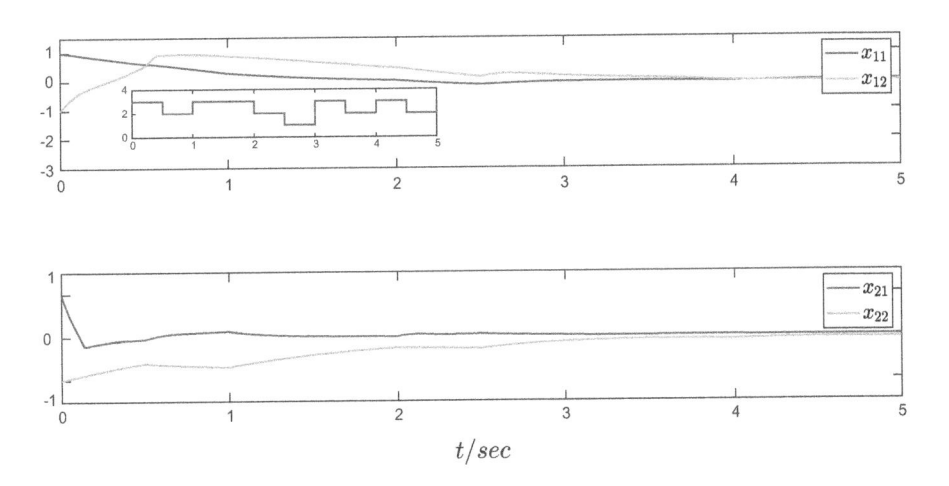

FIGURE 6.3 State response of closed-loop system.

results are presented in Figures 6.3–6.6. Figure 6.3 plots the state response of closed-loop interconnected system under SMC. Figure 6.4 gives the response of sliding surface function for subsystems. Figure 6.5 shows the estimated adaptive parameters $\hat{\beta}_i(t)$. Figure 6.6 depicts the nonlinear dead-zone inputs. In addition, in order to show the superiority of the proposed method, we compare it with traditional state feedback controller under the same control gain matrices. The result is shown in Figure 6.7, from which it is observed that the traditional state feedback controller is not valid.

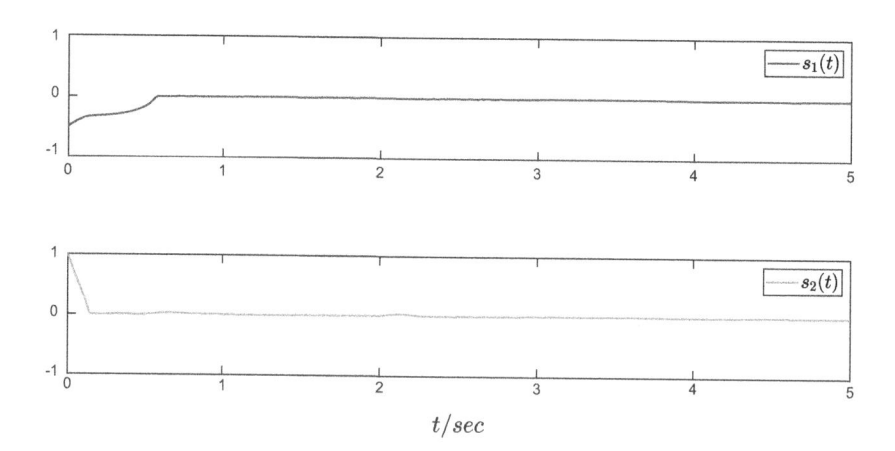

FIGURE 6.4 Sliding surface functions $\hat{\beta}_i(t), i = 1, 2$.

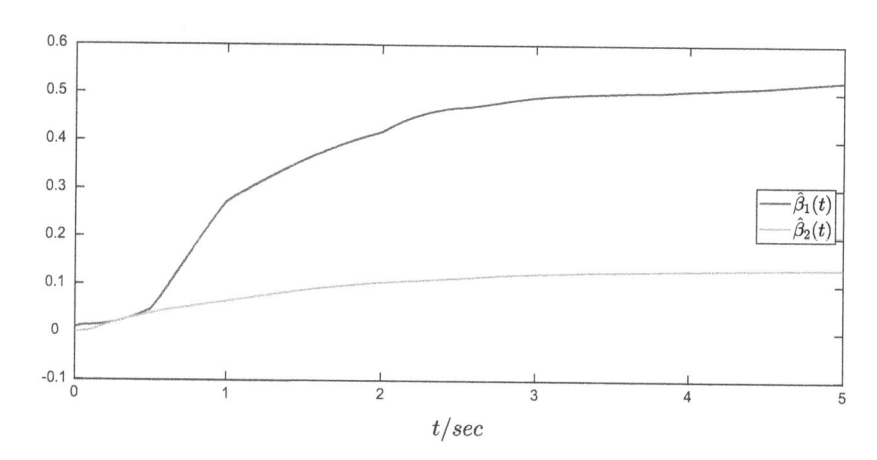

FIGURE 6.5 Estimation values $\hat{\beta}_i(t), i = 1, 2$.

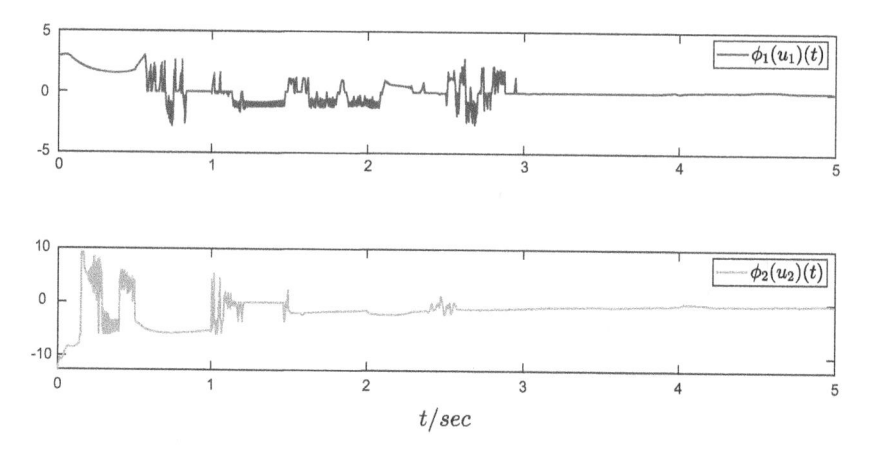

FIGURE 6.6 Nonlinear inputs $\phi_i(u_i), i = 1, 2$.

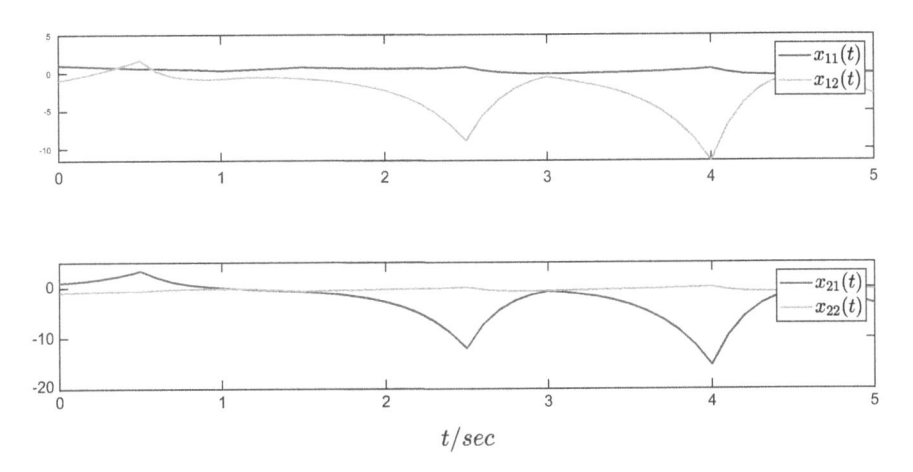

FIGURE 6.7   State response of closed-loop system under traditional state feedback controller.

6.5 CONCLUSION

This chapter has dealt with decentralized adaptive SMC of large-scale semi-Markovian jump-interconnected systems with dead-zone linearity in the input and unknown interconnections among subsystems. Local integral sliding surfaces have been designed for subsystems, based on which local sliding mode dynamics have been obtained. Then, a new method in tackling generally uncertain TRs has been introduced for analyzing stochastic stability of sliding mode dynamics. Further, local adaptive SMC laws have been synthesized to guarantee finite-time reachability of sliding surface and to compensate the effects from dead-zone nonlinearity in the input and unknown interconnections among subsystems. Finally, an example has been proposed to verify the effectiveness of the control scheme.

REFERENCES

[1] J. Xiong, V. A. Ugrinovskii and I. R. Petersen, Local mode dependent decentralized stabilization of uncertain Markovian jump large-scale systems, *IEEE Transactions on Automatic Control*, Vol. 54, No. 11, pp. 2632–2637, 2009.

[2] S. Ma, J. Xiong, V. A. Ugrinovskii and I. R. Petersen, Robust decentralized stabilization of Markovian jump large-scale systems: A neighboring mode dependent control approach, *Automatica*, Vol. 49, No. 10, pp. 3105–3111, 2013.

[3] N. Sandell, P. Varaiya, M. Athans, and M. Safonov, Survey of decentralized control methods for large scale systems, *IEEE Transactions on Automatic Control*, Vol. 23, No. 2, pp. 108–128, 1978.

[4] V. A. Ugrinovskii and H. R. Pota, Decentralized control of power systems via robust control of uncertain Markov jump parameter systems, *International Journal of Control*, Vol. 78, No. 9, pp. 662–677, 2005.

[5] S. Jain and F. Khorrami, Robust decentralized control of power systems utilizing only swing angle measurements, *International Journal of Control*, Vol. 66, No. 4, pp. 581–602, 1997.

[6] S. Elloumi and E. B. Braiek, Robust decentralized control for multimachine power systems-the LMI approach, *Proceedings of IEEE International Conference on Systems Man and Cybernetics* SMC'02, 2002.

[7] G. P. Matthews and R. A. DeCarlo, Decentralized tracking for a class of interconnected nonlinear systems using variable structure control, *Automatica*, Vol. 24, No. 2, pp. 187–193, 1988.

[8] C. H. Chou and C. C. Cheng, A decentralized model reference adaptive variable structure controller for large-scale time-varying delay systems, *IEEE Transactions on Automatic Control*, Vol. 48, No. 7, pp. 1213–1217, 2003.

[9] L. W. Li and G. H. Yang, Decentralised output feedback control of Markovian jump interconnected systems with unknown interconnections, *International Journal of Systems Science*, Vol. 48, No. 9, pp. 1856–1870, 2017.

[10] K. K. Shyu, W. J. Liu and K. C. Hsu, Design of large-scale timedelayed systems with dead-zone input via variable structure control, *Automatica*, Vol. 41, No. 7, pp. 1239–1246, 2005.

[11] V. Utkin, *Sliding Modes in Control Optimization*, Berlin: Springer-Verlag, 1992.

[12] J. Huang and Y. Shi, Stochastic stability of semi-Markov jump linear systems: An LMI approach, *2011 50th IEEE Conference on Decision and Control and European Control Conference. IEEE*, pp. 4668–4673, 2011.

7 Reduced-Order Adaptive Sliding Mode Control for Switching Semi-Markovian Jump Delayed Systems

7.1 INTRODUCTION

In the past years, switched systems have attracted great attention because they not only describe the real mechanism of switching in the system structures, but also for the effectiveness of switching controllers than that of single controller to tackle complex systems. In a switched system [1,2], one of the subsystems (continuous or discrete) is being activated at instant of switching along the state trajectories based on a switching law. Therefore, much investigations on switched systems have been undertaken, theoretically and practically, in which the average dwell time (ADT) switching strategy provides a systematic method to choose a switching law. Based on the ADT approach, many outstanding works have been reported: for instance, the stability analysis problem for a class of switched positive linear systems with ADT switching was investigated in Ref. [3]. The tracking control for a class of switched nonlinear systems in lower triangular form with unknown functions and arbitrary switching was proposed in Ref. [4]; for more details, please refer to Refs. [5–7] and references therein. However, in reality, phenomena like component failure or parameter shifting often exist; then the submodels can potentially be represented by Markovian jump systems (MJSs). Systems of this kind dubbed as switching MJSs that were composed of a series of linear continuous-time or discrete-time subsystems subject to both deterministic switching and stochastic jumping were first proposed by Bolzern et al. [8,9], in which the motivation was to study a failure-prone module and network congestion module. So far, only few research efforts are devoted to study control issues of switching MJSs, such as exponential $l_2 - l_\infty$ stability analysis, finite-time stabilization, passivity and passification [10–12]. Recently, S-MJSs arose great interests among the control community due to their better reflection of behaviors of practical stochastic systems than MJSs. However, the aforementioned efforts were still limited in both theoretical and practical developments with respect to sliding mode control (SMC) of nonlinear systems governed simultaneously by deterministic switching and semi-Markov switching parameters.

Based on the above discussion, this chapter will consider the stabilization of non-linear switching S-MJSs with generally uncertain transition rates (TRs) via adaptive SMC method. Adaptive laws are designed to handle these unknown nonlinearities. Mode-independent common linear switching surface is designed for all submodels such that the potential jumping effect is avoidable. Sliding mode dynamics in low dimensions is obtained based on the linear switching surface. Then, the mean-square exponential stability (MSES) analysis undergoes smoothly because of the simple form of sliding mode dynamics. By designing an adaptive SMC law, the finite-time reachability of switching surface is ensured without requirements on knowledge of the unknown TRs although they are time-varying variables. In the last, our methods are applied to the RLC circuit and DC-DC buck converter with satisfactory results. The main issues to be investigated include the following: (1) a generic method is proposed to deal with dynamical systems with deterministic switchings and stochastic jumps simultaneously; (2) providing the general modeling of TR matrix and less conservative mathematical method in investigating S-MJSs with generally uncertain TRs; (3) proposing reduced-order sliding mode approach in dealing with matched nonlinearities, which leads to the resulting sliding mode dynamics to be totally disturbance-insensitive and order-reduced; (4) designing adaptive SMC laws to ensure finite-time reachability of predefined sliding surface and eliminate system disturbances simultaneously.

7.2   SYSTEM DESCRIPTION

Let's consider the switching semi-Markovian jump delayed system of the form:

$$\begin{cases} \dot{x}(t) = A(g_t, r_t)x(t) + A_d(g_t, r_t)x(t-d) + B(g_t, r_t)[u(t) + f(x(t), x(t-d), t)] \\ x(t+\theta) = \varphi(\theta), \quad \forall \theta \in [-d, 0], \end{cases} \tag{7.1}$$

in which $x(t) \in \mathbb{R}^n$ is the system state, $u(t) \in \mathbb{R}^p$ is the control input, and $\varphi(\theta)$ is an admissible initial condition. $A(g_t, r_t)$, $A_d(g_t, r_t)$ and $B(g_t, r_t)$ are the system real matrices. Particularly, $B(g_t, r_t)$ are considered to have full column rank. d is the system constant time delay. $f(x(t), x(t-d), t)$ is the system nonlinear disturbance constrained by

$$\| f(x(t), x(t-d), t) \| \le l_1 \| x(t) \| + l_2 \| x(t-d) \|,$$

where l_1 and l_2 are unknown real scalars. In the sequel, each value of g_t and r_t will be denoted by α and i, respectively, and $A(g_t, r_t)$ is denoted by $A_{\alpha, i}$, etc.

Remark 7.1

Linear systems were widely investigated in the past few decades, but for complexity of system internal structures and external disturbances, they often show property of nonlinearity. Methods like linearization or constraint by Lipschitz condition were the most commonly used to tackle systems of this kind. In this note, the nonlinearity term $f(x(t), x(t-d), t)$ is linearized by gain constants l_1 and l_2, but the values of which are not known, so we will synthesize adaptive laws to track them in the following, which is more reasonable than by setting the values of l_1 and l_2 known in practice.

Here, g_t is a deterministic switching that takes values in the set $\mathcal{S}_1 = \{1, 2, \ldots, N_1\}$; r_t is a semi-Markov switching that takes values in the set $\mathcal{S}_2 = \{1, 2, \ldots, N_2\}$ and follows the probability transition:

$$\mathbf{Pr}\{r_{t+h} = j \mid r_t = i, g_t = \alpha\} = \begin{cases} \pi_{ij}^{\alpha}(h)h + o(h), & i \neq j, \\ 1 + \pi_{ii}^{\alpha}(h)h + o(h), i = j, \end{cases} \tag{7.2}$$

where $h > 0$ and $\lim_{h \to 0} o(h)/h = 0$, $\pi_{ij}^{\alpha}(h) > 0 (i \neq j)$ is the TR from mode i at time t to mode j at time $t+h$, and $\pi_{ii}^{\alpha}(h) = -\sum_{j \neq i} \pi_{ij}^{\alpha}(h) < 0$ for each $i \in \mathcal{S}_2$.

Also, it is considered that $\pi_{ij}^{\alpha}(h)$ is an uncertain variable in the jumping process, which is not easy to tackle directly in the following part. So, in the following, the presence of $\pi_{ij}^{\alpha}(h)$ meets the two cases: one is that $\pi_{ij}^{\alpha}(h)$ is completely unknown; the other is that $\pi_{ij}^{\alpha}(h)$ is not exactly known but upper- and lower-bounded: for instance, $\pi_{ij}^{\alpha}(h) \in [\underline{\pi}_{ij}^{\alpha}, \overline{\pi}_{ij}^{\alpha}]$, in which $\underline{\pi}_{ij}^{\alpha}$ and $\overline{\pi}_{ij}^{\alpha}$ – the known real constants – are, respectively, the lower and upper bounds of $\pi_{ij}^{\alpha}(h)$. In view of this, we further denote $\pi_{ij}^{\alpha}(h) \triangleq \pi_{ij}^{\alpha} + \Delta\pi_{ij}^{\alpha}(h)$, in which $\pi_{ij}^{\alpha} = \frac{1}{2}(\underline{\pi}_{ij}^{\alpha} + \overline{\pi}_{ij}^{\alpha})$ and $|\Delta\pi_{ij}^{\alpha}(h)| \leq \lambda_{ij}^{\alpha}$ with $\lambda_{ij}^{\alpha} = \frac{1}{2}(\overline{\pi}_{ij}^{\alpha} - \underline{\pi}_{ij}^{\alpha})$. So the TR matrix with $N_2 = 3$ may be described as

$$\Pi^{\alpha} = \begin{bmatrix} \pi_{11}^{\alpha} + \Delta\pi_{11}^{\alpha}(h) & \nabla_{12}^{\alpha} & \pi_{13}^{\alpha} + \Delta\pi_{13}^{\alpha}(h) \\ \nabla_{21}^{\alpha} & \nabla_{22}^{\alpha} & \pi_{23}^{\alpha} + \Delta\pi_{23}^{\alpha}(h) \\ \nabla_{31}^{\alpha} & \pi_{32}^{\alpha} + \Delta\pi_{32}^{\alpha}(h) & \nabla_{33}^{\alpha} \end{bmatrix}, \tag{7.3}$$

where ∇_{ij}^{α} is the description of unknown TRs. For brevity, for $\forall \alpha \in \mathcal{S}_1$, $i \in \mathcal{S}_2$, let $I_i^{\alpha} = I_{i,k}^{\alpha} \cup I_{i,uk}^{\alpha}$, where

$$I_{i,k}^{\alpha} \triangleq \{j : \pi_{ij}^{\alpha} \text{ can be determined for } j \in \mathcal{S}_2\},$$

$$I_{i,uk}^{\alpha} \triangleq \{j : \pi_{ij}^{\alpha} \text{ is not known for } j \in \mathcal{S}_2\}.$$

Here, it is assumed that both $I_{i,k}^{\alpha} \neq \varnothing$ and $I_{i,uk}^{\alpha} \neq \varnothing$. Thus, we can denote the following set:

$$I_{i,k}^{\alpha} \triangleq \{k_{i,1}^{\alpha}, k_{i,2}^{\alpha}, \ldots, k_{i,m}^{\alpha}\} \quad 1 \leq m < N_2,$$

where $k_{i,s}^{\alpha}(s \in \{1, 2, \ldots, m\})$ denotes the index of s-th TR in the i-th row of Π^{α}.

First, we will turn the system (7.1) into a regular form. Since $B_{\alpha,i}$ has full column rank, it is assumed that $B_{\alpha,i} \triangleq \begin{bmatrix} B_{1\alpha,i} \\ B_{2\alpha,i} \end{bmatrix}$ with $B_{1\alpha,i} \in \mathcal{R}^{(n-m)\times m}$ and $B_{2\alpha,i} \in \mathcal{R}^{m\times m}$ being full column rank. Through transformation, $z(t) = T_{\alpha,i}^{-1}x(t)$ with $T_{\alpha,i}^{-1} = \begin{bmatrix} I_{n-m} & -B_{1\alpha,i}B_{2\alpha,i}^{-1} \\ 0 & I_m \end{bmatrix}$. The system (7.1) is equivalent to the form:

$$\dot{z}(t) = \bar{A}_{\alpha,i}z(t) + \bar{A}_{d\alpha,i}z(t-d) + \begin{bmatrix} 0 \\ B_{2\alpha,i} \end{bmatrix}(u(t) + f(T_{\alpha,i}z(t), T_{\alpha,i}z(t-d), t)), \quad (7.4)$$

where $\bar{A}_{\alpha,i} = T_{\alpha,i}^{-1}A_{\alpha,i}T_{\alpha,i}$ and $\bar{A}_{d\alpha,i} = T_{\alpha,i}^{-1}A_{d\alpha,i}T_{\alpha,i}$.

Further, let $z(t) = [z_1^T(t) \ z_2^T(t)]^T$ with $z_1(t) \in \mathcal{R}^{n-m}$ and $z_2(t) \in \mathcal{R}^m$. Denote

$$\bar{A}_{\alpha,i} = \begin{bmatrix} A_{11i}^{\alpha} & A_{12i}^{\alpha} \\ A_{21i}^{\alpha} & A_{22i}^{\alpha} \end{bmatrix}, \bar{A}_{d\alpha,i} = \begin{bmatrix} A_{d11i}^{\alpha} & A_{d12i}^{\alpha} \\ A_{d21i}^{\alpha} & A_{d22i}^{\alpha} \end{bmatrix}.$$

Then, the system (7.4) is decomposed as:

$$\begin{cases} \dot{z}_1(t) = A_{11i}^{\alpha}z_1(t) + A_{12i}^{\alpha}z_2(t) + A_{d11i}^{\alpha}z_1(t-d) + A_{d12i}^{\alpha}z_2(t-d) \\ \dot{z}_2(t) = A_{21i}^{\alpha}z_1(t) + A_{22i}^{\alpha}z_2(t) + A_{d21i}^{\alpha}z_1(t-d) + A_{d22i}^{\alpha}z_2(t-d) \\ \qquad\quad + B_{2\alpha,i}(u(t) + f(T_{\alpha,i}z(t), T_{\alpha,i}z(t-d), t)). \end{cases} \quad (7.5)$$

Definition 7.1 [13]

Consider g_t has an ADT T_a with a $N_0 > 0$; then, $N_g(t_0, t)$ is the number of switches of g_t in the section (t_0, t) that satisfies $N_g(t_0, t) \le N_0 + \dfrac{t-t_0}{T_a}$ for all $t \ge t_0 \ge 0$, where N_0 is called a chatting bound.

7.3 MAIN RESULTS

7.3.1 SMC LAW SYNTHESIS

In this subsection, the linear switching surface function is constructed as:

$$s(t) = -C_{\alpha}z_1(t) + z_2(t), \quad (7.6)$$

where $C_{\alpha}(\alpha \in \mathcal{S}_1)$ are the parameters to be chosen in advance.

By the VSC theory [14], $s(t) = 0$ is true when the switching surface is reached. Consequently, there is $z_2(t) = C_{\alpha}z_1(t)$, with which we combine the system (7.5) leads to the following reduced-order sliding mode dynamics:

$$\dot{z}_1(t) = A^\alpha_{z,i} z_1(t) + A^\alpha_{dz,i} z_1(t-d), \tag{7.7}$$

where $A^\alpha_{z,i} \triangleq A^\alpha_{11i} + A^\alpha_{12i} C_\alpha$ and $A^\alpha_{dz,i} \triangleq A^\alpha_{d11i} + A^\alpha_{d12i} C_\alpha$. Here, C_α is chosen such that $A^\alpha_{z,i}$ is Hurwitz.

Remark 7.2

By using the equivalent control method, we obtained sliding mode dynamics (7.7), which is linear and order-reduced. The desirable superiorities of SMC are sufficiently embodied: the stability analysis of nonlinear switching S-MJSs (7.1) is turned into analysis of the low-dimensional linear sliding mode dynamics (7.7), which is disturbances rejected.

Definition 7.2 [15]

The dynamics (7.7) is mean-square exponentially stable under initial condition $\phi(t)$ and initial $g_0 \in S_1$, $r_0 \in S_2$, if for given scalars $a > 0$ and $b > 0$, it holds

$$\mathbf{E}\{\| z_1(t,\phi(t_0),g_0,r_0) \|^2\} \le a\mathbf{E}\{\sup_{-d \le t \le 0} \| \phi(t) \|^2\} e^{-b(t-t_0)}$$

for all $t \ge 0$.

7.3.2 MEAN-SQUARE EXPONENTIAL STABILITY ANALYSIS

The theorem below provides feasible conditions for the MSES of the system (7.7) based on generally uncertain TRs in the form of linear matrix inequalities (LMIs).

Theorem 7.1

For certain scalars $\mu > 1$ and $\lambda > 0$, the sliding mode dynamics (7.7) subject to uncertain TRs is stable in accordance with Definition 7.2, if there exist matrices $P_{\alpha,i} > 0$, $Q_{\alpha,i} > 0$, $Q_\alpha > 0$, $Z_\alpha > 0$, $V^\alpha_{ij} > 0$, $W^\alpha_{ij} > 0$, $T^\alpha_{ij} > 0$, $U^\alpha_{ij} > 0$ and $M_{\alpha,i}$ of appropriate dimensions with the conditions hold for all $\alpha \in S_1$, $i \in S_2$ as follows:
If $i \in I^\alpha_{i,k}$, $\forall l \in I^\alpha_{i,uk}$, $I^\alpha_{i,k} \triangleq \{k^\alpha_{i,1}, k^\alpha_{i,2}, \ldots, k^\alpha_{i,m1}\}$,

$$\begin{bmatrix} \mathcal{A}^{11}_{\alpha,i} & P_{\alpha,i} A^\alpha_{dz,i} - M_{\alpha,i} & -M_{\alpha,i} & (A^\alpha_{z,i})^T Z_\alpha & \mathcal{A}^{13}_{\alpha,i} \\ * & -e^{-\lambda d} Q_{\alpha,i} & 0 & (A^\alpha_{dz,i})^T Z_\alpha & 0 \\ * & * & -\dfrac{e^{-\lambda d}}{d} Z_\alpha & 0 & 0 \\ * & * & * & -d^{-1} Z_\alpha & 0 \\ * & * & * & * & \mathcal{A}^{14}_{\alpha,i} \end{bmatrix} < 0, \tag{7.8}$$

$$\begin{bmatrix} \mathcal{Q}_{\alpha,i}^1 & Q_{\alpha,k_{i,1}^\alpha} - Q_{\alpha,l} & \cdots & Q_{\alpha,k_{i,m}^\alpha} - Q_{\alpha,l} \\ * & -T_{ik_{i,1}^\alpha}^\alpha & \cdots & 0 \\ \vdots & \vdots & \ddots & \vdots \\ * & * & * & -T_{ik_{i,m1}^\alpha}^\alpha \end{bmatrix} \leq 0, \qquad (7.9)$$

If $i \in \mathrm{I}_{i,uk}^\alpha$, $\forall l \in \mathrm{I}_{i,uk}^\alpha$, $l \neq i$, $\mathrm{I}_{i,k}^\alpha \triangleq \{k_{i,1}^\alpha, k_{i,2}^\alpha, \ldots, k_{i,m2}^\alpha\}$,

$$P_{\alpha,i} - P_{\alpha,l} \geq 0, \qquad (7.10)$$

$$Q_{\alpha,i} - Q_{\alpha,l} \geq 0, \mathrm{V} \qquad (7.11)$$

$$\begin{bmatrix} \mathcal{A}_{\alpha,i}^{21} & P_{\alpha,i}A_{dz,i}^\alpha - M_{\alpha,i} & -M_{\alpha,i} & (A_{z,i}^\alpha)^T Z_\alpha & \mathcal{A}_{\alpha,i}^{23} \\ * & -e^{-\lambda d}Q_{\alpha,i} & 0 & (A_{dz,i}^\alpha)^T Z_\alpha & 0 \\ * & * & -\dfrac{e^{-\lambda d}}{d}Z_\alpha & 0 & 0 \\ * & * & * & -d^{-1}Z_\alpha & 0 \\ * & * & * & * & \mathcal{A}_{\alpha,i}^{24} \end{bmatrix} < 0, \quad (7.12)$$

$$\begin{bmatrix} \mathcal{Q}_{\alpha,i}^2 & Q_{\alpha,k_{i,1}^\alpha} - Q_{\alpha,l} & \cdots & Q_{\alpha,k_{i,m}^\alpha} - Q_{\alpha,l} \\ * & -U_{ik_{i,1}^\alpha}^\alpha & \cdots & 0 \\ \vdots & \vdots & \ddots & \vdots \\ * & * & * & -U_{ik_{i,m2}^\alpha}^\alpha \end{bmatrix} \leq 0, \qquad (7.13)$$

and

$$P_{\alpha,i} \leq \mu P_{\beta,j}, Q_{\alpha,i} \leq \mu Q_{\beta,j}, Q_\alpha \leq \mu Q_\beta,$$
$$Z_\alpha \leq \mu Z_\beta, \forall \alpha, \beta \in S_1, i, j \in S_2, \alpha \neq \beta, \qquad (7.14)$$

the ADT T_a is set by

$$T_a > T_a^* = \frac{\ln \mu}{\lambda}, \qquad (7.15)$$

where

$$\mathcal{A}_{\alpha,i}^{11} = P_{\alpha,i}A_{z,i}^\alpha + (A_{z,i}^\alpha)^T P_{\alpha,i} + M_{\alpha,i} + M_{\alpha,i}^T + Q_{\alpha,i} + dQ_\alpha + \lambda P_{\alpha,i}$$
$$+ \sum_{j \in \mathrm{I}_{i,k}^\alpha} \left[\frac{(\lambda_{ij}^\alpha)^2}{4} W_{ij}^\alpha + \pi_{ij}^\alpha (P_{\alpha,j} - P_{\alpha,l}) \right],$$

$$\mathcal{A}_{\alpha,i}^{13} = [(P_{\alpha,k_{i,1}^\alpha} - P_{\alpha,l}) \ \cdots \ (P_{\alpha,k_{i,m1}^\alpha} - P_{\alpha,l})],$$

$$\mathcal{A}^{14}_{\alpha,i} = [-W^{\alpha}_{ik^{\alpha}_{i,1}} \quad \cdots \quad -W^{\alpha}_{ik^{\alpha}_{i,m1}}],$$

$$Q^{1}_{\alpha,i} = \sum_{j \in I^{\alpha}_{i,k}} \left[\frac{(\lambda^{\alpha}_{ij})^2}{4} T^{\alpha}_{i,j} + \pi^{\alpha}_{ij}(Q_{\alpha,j} - Q_{\alpha,l}) \right] - Q_{\alpha},$$

$$\mathcal{A}^{21}_{\alpha,i} = P_{\alpha,i}A^{\alpha}_{z,i} + (A^{\alpha}_{z,i})^T P_{\alpha,i} + M_{\alpha,i} + M^{T}_{\alpha,i} + Q_{\alpha,i} + dQ_{\alpha} + \lambda P_{\alpha,i}$$

$$+ \sum_{j \in I^{\alpha}_{i,k}} \left[\frac{(\lambda^{\alpha}_{ij})^2}{4} V^{\alpha}_{ij} + \pi^{\alpha}_{ij}(P_{\alpha,j} - P_{\alpha,l}) \right],$$

$$\mathcal{A}^{23}_{\alpha,i} = [(P_{\alpha,k^{\alpha}_{i,1}} - P_{\alpha,l}) \quad \cdots \quad (P_{\alpha,k^{\alpha}_{i,m2}} - P_{\alpha,l})],$$

$$\mathcal{A}^{24}_{\alpha,i} = [-V^{\alpha}_{ik^{\alpha}_{i,1}} \quad \cdots \quad -V^{\alpha}_{ik^{\alpha}_{i,m2}}],$$

$$Q^{2}_{\alpha,i} = \sum_{j \in I^{\alpha}_{i,k}} \left[\frac{(\lambda^{\alpha}_{ij})^2}{4} U^{\alpha}_{i,j} + \pi^{\alpha}_{ij}(Q_{\alpha,j} - Q_{\alpha,l}) \right] - Q_{\alpha}.$$

Proof: Select the Lyapunov functional:

$$V(z_1(t), g_t, r_t) = \sum_{h=1}^{3} V_h(z_1(t), g_t, r_t), \tag{7.16}$$

where

$$V_1(z_1(t), g_t, r_t) = z_1^T(t)P(g_t, r_t)z_1(t),$$

$$V_2(z_1(t), g_t, r_t) = \int_{t-d}^{t} e^{\lambda(\mu-t)} z_1^T(\mu)Q(g_t, r_t)z_1(\mu)d\mu,$$

$$V_3(z_1(t), g_t, r_t) = \int_{-d}^{0} \int_{t+\theta}^{t} e^{\lambda(\mu-t)} z_1^T(\mu)Q(g_t)z_1(\mu)d\mu d\theta$$

$$+ \int_{-d}^{0} \int_{t+\theta}^{t} e^{\lambda(\mu-t)} \dot{z}_1^T(\mu)Z(g_t)\dot{z}_1(\mu)d\mu d\theta.$$

Let $\mathcal{L}$ be the infinitesimal generator; then according to Ref. [16], we have

$$\mathcal{L}V_1(z_1(t), \alpha, r_t) = \lim_{\delta \to 0} \frac{\mathbf{E}\left[V_1(t+\delta, \alpha, r_{t+\delta}) \mid x(t), \alpha, r_t\right] - V_1(z_1(t), \alpha, r_t)}{\delta}$$

$$= \lim_{\delta \to 0} \frac{1}{\delta} \left[\sum_{j=1, j \neq i}^{s} \Pr\{r_{t+\delta} = j \mid \alpha, r_t = i\} z_{\delta}^T P_{\alpha,j} z_{\delta} + \right.$$

$$\left. \Pr\{r_{t+\delta} = j \mid \alpha, r_t = i\} z_{\delta}^T P_{\alpha,i} z_{\delta} - z_1^T(t)P_{\alpha,i}z_1(t) \right], \tag{7.17}$$

where $z_\delta \triangleq z_1(t+\delta)$. Following the conditional probability formula combining a general distribution with the sojourn time in a mode without memoryless property, we have

$$\mathcal{L}V_1(z_1(t),\alpha,r_t) = \lim_{\delta \to 0} \frac{1}{\Delta} \left[\sum_{j=1,j\neq i}^{s} \frac{q_{ij}^\alpha (G_i^\alpha(h+\delta)-G_i^\alpha(t))}{1-G_i^\alpha(h)} z_\delta^T P_{\alpha,j} z_\delta \right.$$

$$\left. + \frac{1-G_i^\alpha(h+\delta)}{1-G_i^\alpha(h)} z_\delta^T P_{\alpha,i} z_\delta - z_1^T(t) P_{\alpha,i} z_1(t) \right]$$

$$= \lim_{\delta \to 0} \frac{1}{\delta} \left[\sum_{j=1,j\neq i}^{s} \frac{q_{ij}^\alpha (G_i^\alpha(h+\delta)-G_i^\alpha(h))}{1-G_i^\alpha(h)} z_\delta^T P_{\alpha,j} z_\delta \right.$$

$$+ \frac{1-G_i^\alpha(h+\delta)}{1-G_i^\alpha(h)} [z_\delta^T - z_1^T(t)] P_{\alpha,i} z_\delta$$

$$+ \frac{1-G_i^\alpha(h+\delta)}{1-G_i^\alpha(h)} z_1^T(t) P_{\alpha,i}^T [z_\delta - z_1(t)]$$

$$\left. - \frac{G_i^\alpha(h+\delta)-G_i^\alpha(h)}{1-G_i^\alpha(h)} z_1^T(t) P_{\alpha,i} z_1(t) \right], \tag{7.18}$$

where q_{ij}^α and $G_i^\alpha(h)$ are the probability intensity and the cumulative distribution function of holding time, respectively. Additionally,

$$\lim_{\delta \to 0} \frac{(G_i^\alpha(h+\delta)-G_i^\alpha(h))}{(1-G_i^\alpha(h))\delta} = \pi_i^\alpha(h), \lim_{\delta \to 0} \frac{1-G_i^\alpha(h+\delta)}{1-G_i^\alpha(h)} = 1,$$

where $\pi_i^\alpha(h)$ denotes the one-step TR from mode i. So

$$\mathcal{L}V_1(z_1(t),\alpha,r_t) = z_1^T(t) \sum_{j\neq i} q_{ij}^\alpha \pi_i^\alpha(h) P_{\alpha,j} z_1(t) + 2z_1^T(t) P_{\alpha,i} \dot{z}_1(t)$$

$$- \pi_i^\alpha(h) z_1^T(t) P_{\alpha,i} z_1(t). \tag{7.19}$$

Now, as in Ref. [16], define $\pi_{ij}^\alpha(h) \triangleq q_{ij} \pi_i^\alpha(h)$ for $j \neq i$ and $\pi_{ii}^\alpha(h) = -\sum_{j\neq i} \pi_{ij}^\alpha(h)$. It further leads to

$$\mathcal{L}V_1(z_1(t),\alpha,i) = z_1^T(t) \left[(A_{z,i}^\alpha)^T P_{\alpha,i} + P_{\alpha,i}^T A_{z,i}^\alpha + \sum_{j=1}^{N_2} \pi_{ij}^\alpha(h) P_{\alpha,j} \right] z_1(t)$$

$$\tag{7.20}$$

$$+ 2z_1^T(t) P_{\alpha,i} A_{dz,i}^\alpha z_1(t-d).$$

Accordingly,

$$\mathcal{L}V_2(z_1(t),\alpha,r_t) = \lim_{\delta \to 0} \frac{1}{\delta}\left[\sum_{j=1,j\neq i}^{s} \frac{q_{ij}^\alpha(G_i^\alpha(h+\delta)-G_i^\alpha(h))}{1-G_i^\alpha(h)}\int_{t+\delta-d}^{t+\delta} e^{\lambda(\mu-t-\delta)}z_1^T(\mu)Q_{\alpha,j}z_1(\mu)\mathrm{d}\mu\right.$$

$$+\frac{1-G_i^\alpha(h+\delta)}{1-G_i^\alpha(h)}\int_{t+\delta-d}^{t+\delta}(e^{\lambda(s-t-\delta)}-e^{\lambda(\mu-t)})z_1^T(\mu)Q_{\alpha,i}z_1(\mu)\mathrm{d}\mu$$

$$+\frac{1-G_i^\alpha(h+\delta)}{1-G_i^\alpha(h)}(\int_{t+\delta-d}^{t+\delta} e^{\lambda(\mu-t)}z_1^T(\mu)Q_{\alpha,i}z_1(\mu)\mathrm{d}\mu$$

$$-\int_{t-d}^{t} e^{\lambda(\mu-t)}z_1^T(\mu)Q_{\alpha,i}z_1(\mu)\mathrm{d}\mu)$$

$$\left.-\frac{G_i^\alpha(h+\delta)-G_i^\alpha(h)}{1-G_i^\alpha(h)}\int_{t-d}^{t} e^{\lambda(\mu-t)}z_1^T(\mu)Q_{\alpha,i}z_1(\mu)\mathrm{d}\mu\right].$$

Similar to the derivative of $\mathcal{L}V_1(z_1(t),\alpha,r_t)$, we have

$$\mathcal{L}V_2(z_1(t),\alpha,i) = z_1^T(t)Q_{\alpha,i}z_1(t) + \sum_{j=1}^{N_2} \pi_{ij}^\alpha(h)\int_{t-d}^{t} e^{\lambda(\mu-t)}z_1^T(\mu)Q_{\alpha,j}z_1(\mu)\mathrm{d}\mu$$

$$-e^{-\lambda d}z_1^T(t-d)Q_{\alpha,i}z_1(t-d) - \lambda V_2(z_1(t),\alpha,i). \tag{7.21}$$

Overall, we have

$$\mathcal{L}V(z_1(t),\alpha,i) = z_1^T(t)[P_{\alpha,i}A_{z,i}^\alpha + (A_{z,i}^\alpha)^T P_{\alpha,i} + Q_{\alpha,i} + dQ_\alpha + \sum_{j=1}^{N_2}\pi_{ij}^\alpha(h)P_{\alpha,j}$$

$$+\lambda P_{\alpha,i}]z_1(t) + 2z_1^T(t)P_{\alpha,i}A_{dz,i}^\alpha z_1(t-d)$$

$$-e^{-\lambda d}z_1^T(t-d)Q_{\alpha,i}z_1(t-d) - \lambda V(z_1(t),\alpha,i)$$

$$+\int_{t-d}^{t} e^{\lambda(\mu-t)}z_1^T(\mu)\left[\sum_{j=1}^{N_2}\pi_{ij}^\alpha(h)Q_{\alpha,j} - Q_\alpha\right]z_1(\mu)\mathrm{d}\mu$$

$$-\int_{t-d}^{t} e^{\lambda(\mu-t)}\dot{z}_1^T(\mu)Z_\alpha\dot{z}_1(\mu)\mathrm{d}\mu + d\dot{z}_1^T(t)Z_\alpha\dot{z}_1(t) \tag{7.22}$$

in which we denote $\sum_{j=1}^{N_2}\pi_{ij}^\alpha(h)Q_{\alpha,j} - Q_\alpha \leq 0$. Further, it holds that

$$-\int_{t-d}^{t} e^{\lambda(\mu-t)}\dot{z}_1^T(\mu)Z_\alpha\dot{z}_1(\mu)\mathrm{d}\mu \leq -\frac{e^{-\lambda d}}{d}\left\{\int_{t-d}^{t}\dot{z}_1(\mu)d\mu\right\}^T Z_\alpha\left\{\int_{t-d}^{t}\dot{z}_1(\mu)d\mu\right\}. \tag{7.23}$$

Following Newton–Leibniz formula, for given free-weighting matrices $M_{\alpha,i}$, it holds

$$2z_1^T(t)M_{\alpha,i}\left[z_1(t)-z_1(t-d)-\int_{t-d}^t \dot{z}_1(\mu)d\mu\right]=0. \tag{7.24}$$

Combining (7.20)–(7.24) leads to

$$\mathcal{L}V(z_1(t),\alpha,i)+\lambda V(z_1(t),\alpha,i)\leq \eta^T(t)\left(\Gamma_i^\alpha+diag\left\{\sum_{j=1}^{N_2}\pi_{ij}^\alpha(h)P_{\alpha,j},0,0\right\}\right)\eta(t), \tag{7.25}$$

where $\eta(t)=\left[z_1^T(t)\ \ z_1^T(t-d)\ \left(\int_{t-d}^t \dot{z}_1(\mu)d\mu\right)^T\right]^T$,

$$\Gamma_i^\alpha=\begin{bmatrix}\Gamma_{1i}^\alpha & P_{\alpha,i}^T A_{d\alpha,i}-M_{\alpha,i} & -M_{\alpha,i}\\ * & -e^{-\lambda d}Q_{\alpha,i} & 0\\ * & * & -\dfrac{e^{-\lambda d}}{d}Z_\alpha\end{bmatrix}+\begin{bmatrix}(A_{z,i}^\alpha)^T Z_\alpha\\ (A_{dz,i}^\alpha)^T Z_\alpha\\ 0\end{bmatrix}dZ_\alpha^{-1}\begin{bmatrix}(A_{z,i}^\alpha)^T Z_\alpha\\ (A_{dz,i}^\alpha)^T Z_\alpha\\ 0\end{bmatrix}^T$$

with $\Gamma_{1i}^\alpha=(A_{z,i}^\alpha)^T P_{\alpha,i}+P_{\alpha,i}A_{z,i}^\alpha+M_{\alpha,i}+M_{\alpha,i}^T+Q_{\alpha,i}+dQ_\alpha+\lambda P_{\alpha,i}$.

So, if $\hat{\Gamma}_i^\alpha\triangleq\Gamma_i^\alpha+diag\left\{\sum_{j=1}^{N_2}\pi_{ij}^\alpha(h)P_{\alpha,j},0,0\right\}<0$, then from (7.25), we know that

$$\mathcal{L}V(z_1(t),\alpha,i)+\lambda V(z_1(t),\alpha,i)<0. \tag{7.26}$$

Multiplying (7.26) by $e^{\lambda t}$ and then integrating both sides of (7.26) between the interval $t\in[t_k,t_{k+1}]$ yield

$$\mathbf{E}\{V(z_1(t),r(t),g(t))\}\leq e^{-\lambda(t-t_k)}\mathbf{E}\{V(z_1(t_k),r(t),g(t))\}. \tag{7.27}$$

Combining (7.14) and (7.16), we know the following inequality holds at switching instant t_k,

$$\mathbf{E}\{V(z_1(t),r(t),g(t))\}\leq \mu\mathbf{E}\{V(z_1(t_k^-),r(t),g(t))\}. \tag{7.28}$$

Therefore, it follows from (7.27), (7.28) and $k=N_g(t_0,t)\leq(t-t_0)/T_\alpha$ that

$$\mathbf{E}\{V(z_1(t),r(t),g(t))\}\leq \mu e^{-\lambda(t-t_k)}\mathbf{E}\{V(z_1(t_k^-),r(t),g(t))\}$$

$$\leq \mu^2 e^{-\lambda(t-t_k)}\mathbf{E}\{V(z_1(t_{k-1}),r(t),g(t))\}$$

$$\leq\ldots\leq \mu^k e^{-\lambda(t-t_k)}V(z_1(t_0),r_0,g_0) \tag{7.29}$$

$$\leq e^{-(\lambda-\frac{\ln\mu}{T_a})(t-t_0)}V(z_1(t_0),r_0,g_0).$$

Recalling (7.16), there exists a scalar ϵ_1 satisfying $0 < \epsilon_1 \le \min\limits_{\alpha \in S_1, i \in S_2}(\lambda_{min}(P_{\alpha,i}))$ such that

$$V(z_1(t), r(t), g(t)) \ge \epsilon_1 \| z_1(t) \|^2 . \tag{7.30}$$

Furthermore, denoting $\epsilon_2 = \max\limits_{\alpha \in S_1, i \in S_2}(\lambda_{max}(P_{\alpha,i}))$, $\epsilon_3 = \max\limits_{\alpha \in S_1, i \in S_2}(\lambda_{max}(Q_{\alpha,i}))$, $\epsilon_4 = \max\limits_{\alpha \in S_1}(\lambda_{max}(Q_\alpha))$, $\epsilon_5 = \max\limits_{\alpha \in S_1}(\lambda_{max}(Z_\alpha))$ yields

$$V(z_1(0), t_0, r_0, g_0) \le \epsilon_2 \| z_1(t_0) \|^2 + \epsilon_3 \int_{t_0-d}^{t_0} e^{\lambda(\mu - t_0)} \| z_1(\mu) \|^2 \, d\mu$$

$$+ \epsilon_4 \int_{-d}^{0} \int_{t_0+\theta}^{t_0} e^{\lambda(\mu - t_0)} (\epsilon_4 \| z_1(\mu) \|^2 + \epsilon_5 \| \dot{z}_1(\mu) \|^2) d\mu d\theta. \tag{7.31}$$

Combining (7.29)–(7.31) derives $\| z_1(t) \|^2 \le a e^{-b(t-t_0)} \| z_1(t_0) \|^2$, in which

$$a = \frac{\epsilon_2 + \dfrac{\epsilon_3}{\lambda}(1 - e^{-\lambda d}) + (\epsilon_4 + \epsilon_5)\dfrac{1}{\lambda^2}[\lambda d - 1 + e^{-\lambda d}]}{\epsilon_1}, \quad b = \lambda - \frac{\ln \mu}{T_a}, \quad \text{which means the}$$

MSES of sliding mode dynamics (7.7) is proved according to Definition 7.2. Since we consider a generally uncertain TRs in this paper, the following two cases have to take into consideration:

Case I $i \in I_{i,k}^\alpha$.

First, let $\lambda_{i,k}^\alpha \triangleq \sum\limits_{j \in I_{i,k}^\alpha} \pi_{ij}^\alpha(h)$. Since $I_{i,uk}^\alpha \ne \varnothing$, it holds that $\lambda_{i,k}^\alpha < 0$. Notice that $\sum\limits_{j=1}^{N_2} \pi_{ij}^\alpha(h) P_{\alpha,j}$ can be represented as

$$\sum_{j=1}^{N_2} \pi_{ij}^\alpha(h) P_{\alpha,j} = \left(\sum_{j \in I_{i,k}^\alpha} + \sum_{j \in I_{i,uk}^\alpha} \right) \pi_{ij}^\alpha(h) P_{\alpha,j}$$

$$= \sum_{j \in I_{i,k}^\alpha} \pi_{ij}^\alpha(h) P_{\alpha,j} - \lambda_{i,k}^\alpha \sum_{j \in I_{i,uk}^\alpha} \frac{\pi_{ij}^\alpha(h)}{-\lambda_{i,k}^\alpha} P_{\alpha,j}, \tag{7.32}$$

and it is obvious that $0 \le \pi_{ij}^\alpha(h) / -\lambda_{i,k}^\alpha \le 1 \; (j \in I_{i,uk}^\alpha)$ and $\sum\limits_{j \in I_{i,uk}^\alpha} \dfrac{\pi_{ij}^\alpha(h)}{-\lambda_{i,k}^\alpha} = 1$. So for $\forall l \in I_{i,uk}^\alpha$, there is

$$\hat{\Gamma}_i^\alpha = \sum_{j \in I_{i,uk}^\alpha} \frac{\pi_{ij}^\alpha(h)}{-\lambda_{i,k}^\alpha} \left[\Gamma_i^\alpha + diag\{ \sum_{j \in I_{i,k}^\alpha} \pi_{ij}^\alpha(h)(P_{\alpha,j} - P_{\alpha,l}), 0, 0 \} \right]. \tag{7.33}$$

Therefore, for $0 \le \pi_{ij}^{\alpha}(h) \le -\lambda_{i,k}^{\alpha}$, $\hat{\Gamma}_i^{\alpha} < 0$ is equivalent to

$$\Gamma_i^{\alpha} + diag\{\sum_{j \in I_{i,k}^{\alpha}} \pi_{ij}^{\alpha}(h)(P_{\alpha,j} - P_{\alpha,l}), 0, 0\} < 0. \tag{7.34}$$

In formula (7.34), it is true that

$$\sum_{j \in I_{i,k}^{\alpha}} \pi_{ij}^{\alpha}(h)(P_{\alpha,j} - P_{\alpha,l}) = \sum_{j \in I_{i,k}^{\alpha}} \pi_{ij}^{\alpha}(P_{\alpha,j} - P_{\alpha,l}) + \sum_{j \in I_{i,k}^{\alpha}} \Delta\pi_{ij}^{\alpha}(h)(P_{\alpha,j} - P_{\alpha,l}), \tag{7.35}$$

then by virtue of Lemma 1.4 and for any $W_{ij}^{\alpha} > 0$, it follows that

$$\sum_{j \in I_{i,k}^{\alpha}} \Delta\pi_{ij}^{\alpha}(h)(P_{\alpha,j} - P_{\alpha,l}) = \sum_{j \in I_{i,k}^{\alpha}} \left[\frac{1}{2} \Delta\pi_{ij}^{\alpha}(h)(P_{\alpha,j} - P_{\alpha,l}) + (P_{\alpha,j} - P_{\alpha,l})) \right]$$

$$\le \sum_{j \in I_{i,k}^{\alpha}} \left[\frac{(\lambda_{ij}^{\alpha})^2}{4} W_{ij}^{\alpha} + (P_{\alpha,j} - P_{\alpha,l})(W_{ij}^{\alpha})^{-1}(P_{\alpha,j} - P_{\alpha,l})^T \right]. \tag{7.36}$$

From (7.32)–(7.36), by applying Schur complement, we can see that (7.8) guarantees $\Gamma_i^{\alpha} < 0$ when $i \in I_{i,k}^{\alpha}$. Similarly, $\sum_{j=1}^{N_2} \pi_{ij}^{\alpha}(h)Q_{\alpha,j} - Q_{\alpha} \le 0$ is guaranteed by (7.9).

Case II $i \in I_{i,uk}^{\alpha}$.

Similarly, denote $\lambda_{i,k}^{\alpha} \triangleq \sum_{j \in I_{i,k}^{\alpha}} \pi_{ij}^{\alpha}(h)$. Since $I_{i,k}^{\alpha} \ne \varnothing$, it holds that $\lambda_{i,k}^{\alpha} > 0$. Now, $\sum_{j=1}^{N_2} \pi_{ij}^{\alpha}(h)P_{\alpha,j}$ can be represented as

$$\sum_{j=1}^{N_2} \pi_{ij}^{\alpha}(h)P_{\alpha,j} = \sum_{j \in I_{i,k}^{\alpha}} \pi_{ij}^{\alpha}(h)P_{\alpha,j} + \pi_{ii}^{\alpha}(h)P_{\alpha,i} + \sum_{j \in I_{i,uk}^{\alpha}, j \ne i} \pi_{ij}^{\alpha}(h)P_{\alpha,j}$$

$$= \sum_{j \in I_{i,k}^{\alpha}} \pi_{ij}^{\alpha}(h)P_{\alpha,j} + \pi_{ii}^{\alpha}(h)P_{\alpha,i}$$

$$- (\pi_{ii}^{\alpha}(h) + \lambda_{i,k}^{\alpha}) \sum_{j \in I_{i,uk}^{\alpha}, j \ne i} \frac{\pi_{ij}^{\alpha}(h)P_{\alpha,j}}{-\pi_{ii}^{\alpha}(h) - \lambda_{i,k}^{\alpha}}, \tag{7.37}$$

and   it   is   obvious   that $0 \le \pi_{ij}^{\alpha}(h) / -\pi_{ii}^{\alpha}(h) - \lambda_{i,k}^{\alpha} \le 1$ $(j \in I_{i,uk}^{\alpha})$ and $\sum_{j \in I_{i,uk}^{\alpha}, j \ne i} \frac{\pi_{ij}^{\alpha}(h)}{-\pi_{ii}^{\alpha}(h) - \lambda_{i,k}^{\alpha}} = 1$. So for $\forall l \in I_{i,uk}^{\alpha}, l \ne i$,

$$\hat{\Gamma}_i^\alpha = \sum_{j \in I_{i,uk}^\alpha, j \neq i} \frac{\pi_{ij}^\alpha(h)}{-\pi_{ii}^\alpha(h) - \lambda_{i,k}^\alpha} \left[\Gamma_i^\alpha + diag \left\{ \pi_{ii}^\alpha(h)(P_{\alpha,i} - P_{\alpha,l}) \right.\right.$$

$$\left.\left. + \sum_{j \in I_{i,k}^\alpha} \pi_{ij}^\alpha(h)(P_{\alpha,j} - P_{\alpha,l}), 0, 0 \right\} \right]. \tag{7.38}$$

Therefore, for $0 \le \pi_{ij}^\alpha(h) \le -\pi_{ii}^\alpha(h) - \lambda_{i,k}^\alpha$, $\Gamma_i^\alpha < 0$ is equivalent to

$$\Gamma_i^\alpha + diag \left\{ \pi_{ii}^\alpha(h)E^T(P_{\alpha,i} - P_{\alpha,l}) + \sum_{j \in I_{i,k}^\alpha} \pi_{ij}^\alpha(h)(P_{\alpha,j} - P_{\alpha,l}), 0, 0 \right\} < 0. \tag{7.39}$$

Since $\pi_{ii}^\alpha(h) < 0$, (7.39) holds if we have

$$\begin{cases} P_{\alpha,i} - P_{\alpha,l} \ge 0, \\ \\ \Gamma_i^\alpha + diag \left\{ \sum_{j \in I_{i,k}^\alpha} \pi_{ij}^\alpha(h)(P_{\alpha,j} - P_{\alpha,l}), 0, 0 \right\} < 0. \end{cases} \tag{7.40}$$

Also, as in (7.35) and (7.36), for any $V_{ij}^\alpha > 0$, we have

$$\sum_{j \in I_{i,k}^\alpha} \pi_{ij}^\alpha(h)(P_{\alpha,j} - P_{\alpha,l}) \le \sum_{j \in I_{i,k}^\alpha} \pi_{ij}^\alpha(P_{\alpha,j} - P_{\alpha,l}) + \sum_{j \in I_{i,k}^\alpha} \left[\frac{(\lambda_{ij}^\alpha)^2}{4} V_{ij}^\alpha \right.$$

$$\left. + (P_{\alpha,j} - P_{\alpha,l})(V_{ij}^\alpha)^{-1}(P_{\alpha,j} - P_{\alpha,l})^T \right]. \tag{7.41}$$

From (7.37)–(7.41), we know that (7.10) and (7.12) guarantee $\Gamma_i^\alpha < 0$ by applying Schur complement when $i \in I_{i,uk}^\alpha$. To deal with $\sum_{j=1}^{N_2} \pi_{ij}^\alpha(h)Q_{\alpha,j} - Q_\alpha \le 0$, we can follow (7.32)–(7.41). In summary, the MSES of the system (7.7) with generally uncertain TRs is proved from the above analysis. This completes the proof.

7.3.3 Adaptive SMC Law Design

In this part, an adaptive SMC law is constructed so that state reachability of sliding surface can be satisfied before the sliding motion. The detailed analysis is provided in the following theorem to ensure reachability of the switching surface $s(t) = 0$.

Noting that the system nonlinear estimation bounds l_1 and l_2 in nonlinear item are not known in the real system. Therefore, $\hat{l}_1(t)$ and $\hat{l}_2(t)$ are used to adapt them and the corresponding estimation errors are $\tilde{l}_1(t) = \hat{l}_1(t) - l_1$ and $\tilde{l}_2(t) = \hat{l}_2(t) - l_2$, respectively.

Theorem 7.2

By designing the switching surface function (7.6), suppose the conditions in Theorem 7.1 have feasible solutions. Then, the following adaptive sliding mode law can force the state of controlled system onto the switching surface $s(t) = 0$ by finite time:

$$\begin{cases} u(t) = -B_{2\alpha,i}^{-1}[u_1(t) + u_2(t)], \\ u_1(t) = \bar{C}_\alpha[A_{\alpha,i}z(t) + A_{d\alpha,i}z(t-d)], \\ u_2(t) = [\rho + \| T_{\alpha,i}z(t) \| \hat{l}_1(t) + \| T_{\alpha,i}z(t-d) \| \hat{l}_2(t)]sgn(s(t)), \end{cases} \tag{7.42}$$

where $\bar{C}_\alpha = [-C_\alpha \quad I]$ and $\rho > 0$ is a small constant. In addition, the adaptive gains are designed as

$$\dot{\hat{l}}_1(t) = c_1 \| B_{2\alpha,i} \| \| T_{\alpha,i}z(t) \| \| s(t) \|,$$

$$\dot{\hat{l}}_2(t) = c_2 \| B_{2\alpha,i} \| \| T_{\alpha,i}z(t-d) \| \| s(t) \|,$$

where $c_i > 0 (i = 1,2)$ are the desired scalars.

Proof: Select the Lyapunov functional candidate:

$$V(t) = 0.5[s^T(t)s(t) + c_1^{-1}\tilde{l}_1^2(t) + c_2^{-1}\tilde{l}_2^2(t)]. \tag{7.43}$$

Then, it gives

$$\begin{aligned} \mathcal{L}V(t) &= s^T(t)\dot{s}(t) + c_1^{-1}\tilde{l}_1(t)\dot{\tilde{l}}_1(t) + c_2^{-1}\tilde{l}_2(t)\dot{\tilde{l}}_2(t) \\ &= s^T(t)\bar{C}_\alpha(A_{\alpha,i}z(t) + A_{d\alpha,i}z(t-d)) + s^T(t)B_{2\alpha,i}(u(t) \\ &\quad + f(T_{\alpha,i}z(t), T_{\alpha,i}z(t-d),t) + c_1^{-1}\tilde{l}_1^2(t) + c_2^{-1}\tilde{l}_2^2(t). \end{aligned} \tag{7.44}$$

Substituting (7.42) into (7.44) yields

$$\begin{aligned} \mathcal{L}V(t) &= -\rho s^T(t)sgn(s(t)) - [\| T_{\alpha,i}z(t) \| \hat{l}_1(t) + \| T_{\alpha,i}z(t-d) \| \hat{l}_2(t)]s^T(t)sgn(s(t) \\ &\quad + s^T(t)B_{2\alpha,i}f(T_{\alpha,i}z(t), T_{\alpha,i}z(t-d),t) + c_1^{-1}\tilde{l}_1(t)\dot{\tilde{l}}_1(t) + c_2^{-1}\tilde{l}_2(t)\dot{\tilde{l}}_2(t) \\ &\leq -\rho \| s(t) \| - [\| T_{\alpha,i}z(t) \| \hat{l}_1(t) + \| T_{\alpha,i}z(t-d) \| \hat{l}_2(t)] \| s(t) \| \\ &\quad + \| s(t) \| \| B_{2\alpha,i} \| (l_1 \| T_{\alpha,i}z(t) \| + l_2 \| T_{\alpha,i}z(t-d) \|) \\ &\quad + c_1^{-1}\tilde{l}_1(t)\dot{\tilde{l}}_1(t) + c_2^{-1}\tilde{l}_2(t)\dot{\tilde{l}}_2(t) \end{aligned} \tag{7.45}$$

Notice that it holds

$$\dot{\tilde{l}}_1(t) = \dot{\hat{l}}_1(t), \dot{\tilde{l}}_1(t) = \dot{\hat{l}}_1(t).$$

Recalling (7.45) and the above qualities, then it follows

$$\mathcal{L}V(t) \leq -\rho \, \| \, s(t) \, \| < 0. \quad \text{(for } s(t) \neq 0\text{)} \tag{7.46}$$

Furthermore, seeing from the regulation laws, $\dot{\tilde{l}}_i(t) = \dot{\hat{l}}_i(t)$, which implies we have an instant T^* that satisfies $\tilde{l}_i(t) > 0$ for $t > T^*$. Then, it obtains $c_1^{-1}\tilde{l}_i(t)\dot{\tilde{l}}_i(t) > 0$. From (7.46), it is seen that reachability condition $s^T(t)\dot{s}(t) < 0$ holds for all $t > T^*$. To this end, the reachability of switching surface $s(t) = 0$ is confirmed, which completes the proof.

Remark 7.3

Inevitably, the shortage of SMC method from the sign function in the control (11) may produce undesired chattering effect. Therefore, the last few decades have witnessed that a lot of methods were proposed to reduce the chattering effect, which include the "quasi-sliding mode control," "reaching law method," "dynamic sliding mode method," etc. In addition, it is seen from the formula (7.42) that the tuning scalar ρ directly determines the reaching time of the sliding surface.

Remark 7.4

As we can see, the overall analysis of the original system via adaptive sliding mode approach can be easily applied to none of the switching signal case, that is, $g_t = \{1\}$, which means a normal S-MJS. Further, it is seen that when $\Delta\pi_{ij}(h) = 0$ in Π^{α}, we have normal MJSs model with generally uncertain TRs, which have been studied in Refs. [17,18].

7.4 NUMERICAL EXAMPLES

Example 7.1

Consider a RLC series circuit in Figure 7.1. Accordingly, let $x_1(t) = u_c(t)$ and $x_2(t) = i_L(t)$ as state vectors, where $u_c(t)$ and $i_L(t)$ are the voltage of the capacitor and the current of the inductor, respectively. So the model in Figure 7.1 can be represented by the system (7.1) with switching signal governed by *Switch* with values through 1 or 2, and *Jumping* between 1, 2 and 3 of capacitors follows semi-Markovian switching. The dynamics is described by:

$$L(g_t)\frac{di_L(t)}{dt} + u_c(t) + R_1 i_L(t) = u(t),$$

$$C(r_t)\frac{du_c(t)}{dt} + \frac{u_c(t)}{R_2} = i_L(t).$$

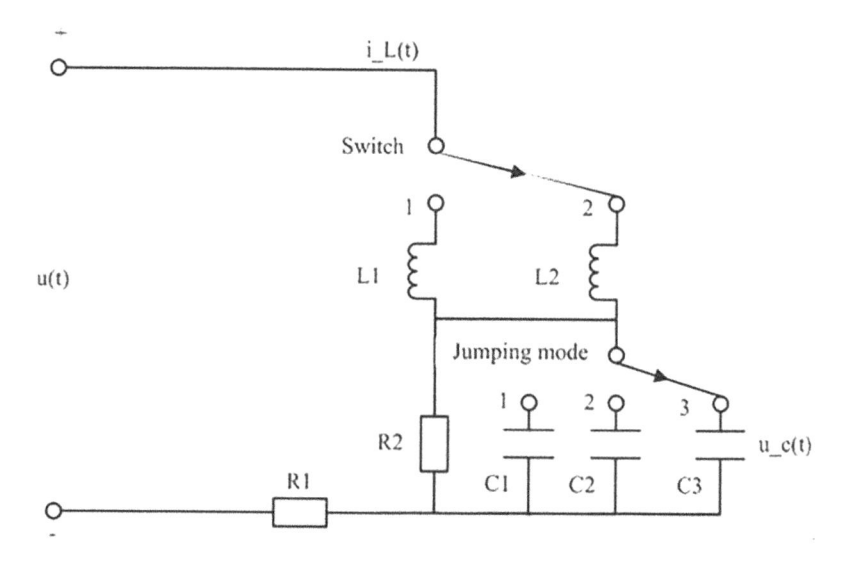

FIGURE 7.1　RLC series circuit.

Let $x(t) = [x_1^T(t) \ x_2^T(t)]^T$; then it obtains state space in the form:

$$\dot{x}(t) = \begin{bmatrix} -\dfrac{1}{C(r_i)R_2} & \dfrac{1}{C(r_i)} \\[2mm] -\dfrac{1}{L(g_i)} & -\dfrac{R_1}{L(g_i)} \end{bmatrix} x(t) + \begin{bmatrix} 0 \\[2mm] \dfrac{1}{L(g_i)} \end{bmatrix} u(t).$$

When the circuit contains time delay $d = 0.1\,\text{s}$ and nonlinear uncertainties, more generality, let's consider the system:

$$\dot{x}(t) = \begin{bmatrix} -\dfrac{1}{C(r_i)R_2} & \dfrac{1}{C(r_i)} \\[2mm] -\dfrac{1}{L(g_i)} & -\dfrac{R_1}{L(g_i)} \end{bmatrix} x(t) + \begin{bmatrix} 0.1 & 0 \\[1mm] 0 & -0.1 \end{bmatrix} x(t-0.1)$$

$$+ \begin{bmatrix} 0 \\[2mm] \dfrac{1}{L(g_i)} \end{bmatrix} (u(t) + f(x(t), x(t-d))),$$

where $L_1 = 1H$, $L_2 = 2H$, $C_1 = 0.5F$, $C_2 = 0.2F$, $C_3 = 0.1F$, $R_1 = 20\Omega$ and $R_2 = 10\Omega$. The nonlinear function is given as $f(x(t), x(t-d), t) = 0.2\sin^2(t)(x_1(t) + x_2(t-0.1))$; therefore, we have the following parameters for the system:

$$A_{1,1} = \begin{bmatrix} -0.2 & 2 \\ -1 & -20 \end{bmatrix}, A_{1,2} = \begin{bmatrix} -0.5 & 5 \\ -1 & -20 \end{bmatrix}, A_{1,3} = \begin{bmatrix} -1 & 10 \\ -1 & -20 \end{bmatrix},$$

$$A_{2,1} = \begin{bmatrix} -0.2 & 2 \\ -0.5 & -10 \end{bmatrix}, A_{2,2} = \begin{bmatrix} -0.5 & 5 \\ -0.5 & -10 \end{bmatrix}, A_{2,3} = \begin{bmatrix} -1 & 10 \\ -0.5 & -10 \end{bmatrix},$$

$$A_{d\alpha,i} = \begin{bmatrix} 0.1 & 0 \\ 0 & -0.1 \end{bmatrix}, (\alpha = 1,2; i = 1,2,3),$$

$$B_{1,1} = B_{1,2} = B_{1,3} = \begin{bmatrix} 0 \\ 1 \end{bmatrix}, B_{2,1} = B_{2,2} = B_{2,3} = \begin{bmatrix} 0 \\ 0.5 \end{bmatrix},$$

The generally uncertain TR matrices are given as follows:

$$\Pi^1 = \begin{bmatrix} -2.8 + \Delta\pi_{11}^1(h) & \nabla_{12}^1 & \nabla_{13}^1 \\ \nabla_{21}^1 & \nabla_{22}^1 & 1.5 + \Delta\pi_{23}^1(h) \\ 1.1 + \Delta\pi_{31}^1(h) & \nabla_{32}^1 & -2.3 + \Delta\pi_{33}^1(h) \end{bmatrix}.$$

$$\Pi^2 = \begin{bmatrix} \nabla_{11}^2 & 1.0 + \Delta\pi_{12}^2(h) & \nabla_{13}^2 \\ 0.8 + \Delta\pi_{21}^2(h) & -1.5 + \Delta\pi_{22}^2(h) & \nabla_{23}^2 \\ \nabla_{31}^2 & \nabla_{32}^2 & -2.5 + \Delta\pi_{33}^2(h) \end{bmatrix}.$$

It can be seen from these parameters that the system is already regular, so we do not need to take linear transformation. By defining $x_1(t) = z_1(t)$ and $x_2(t) = z_2(t)$, let $\Delta\pi_{ij}^\alpha(h) \le \lambda_{ij}^\alpha = |0.1 * \pi_{ij}^\alpha(h)|$, $\mu = 1.2$ and $\lambda = 0.1$. By checking the conditions in Theorem 7.1 with $C_1 = -0.5$ and $C_2 = -0.4$, we obtain the following feasible solutions:

$$P_{11} = 11.0682, P_{12} = 11.7482, P_{13} = 7.4577, P_{21} = 38.2543, P_{22} = 23.6994, P_{23} = 17.8974,$$

$$Q_{11} = 13.7916, Q_{12} = 19.6871, Q_{13} = 21.5970, Q_{21} = 38.8290, Q_{22} = 43.3359, Q_{23} = 33.3265,$$

$$M_{11} = -7.0342, M_{12} = 4.4922, M_{13} = 3.3430, M_{21} = -1.5543, M_{22} = 5.0152, M_{23} = 10.5328,$$

$$Q_1 = 38.9940, Q_2 = 51.4770, Z_1 = 2.7637, Z_2 = 3.7816.$$

By computing (7.15), we have $T_a^* = 1.8232$. Therefore, the ADT is chosen as $T_a = 2$. The initial conditions are given as $\varphi(\theta) = [5 \ -5]^T, \theta \in [-0.1,0]$, the adaptive laws are governed as in Theorem 7.2, and the corresponding parameters are chosen as $c_1 = 0.015$, $c_2 = 0.01$ and $\hat{l}_1(0) = \hat{l}_2(0) = 0$, and the tuning scalar ρ in controller (7.42) is selected as $\rho = 0.01$. In addition, in order to reduce the chattering effect of switching signals, here, $\text{sgn}(s(t))$ is changed by $s(t)/(\|s(t)\| +0.01)$. Then, Figures 7.2–7.5 present simulation results. Figure 7.2 plots the state response of the closed-loop system under one possible deterministic switching and stochastic jumping. In Figure 7.3, the switching surface function $s(t)$ is given. Figure 7.4 demonstrates the estimated values $\hat{l}_1(t)$ and $\hat{l}_2(t)$ in Ref. (11). Figure 7.5 plots the control input.

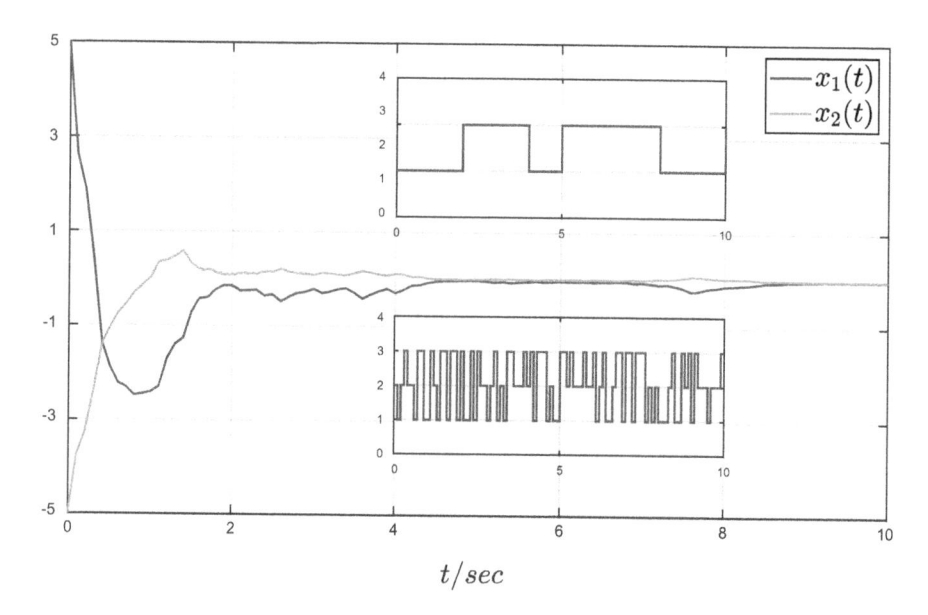

FIGURE 7.2   The time response of system state $x(t)$.

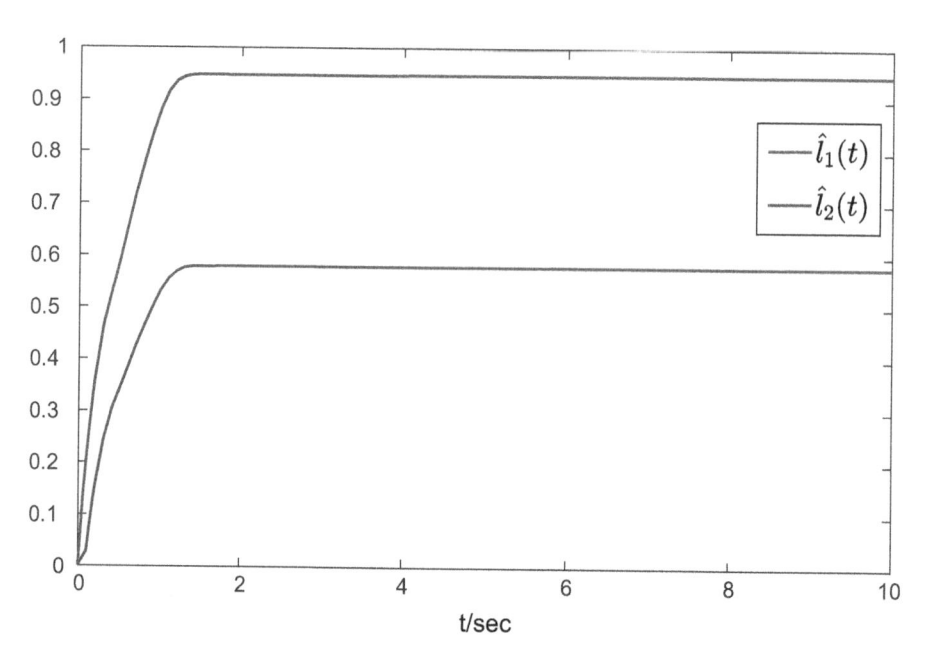

FIGURE 7.3   Adaptive gains.

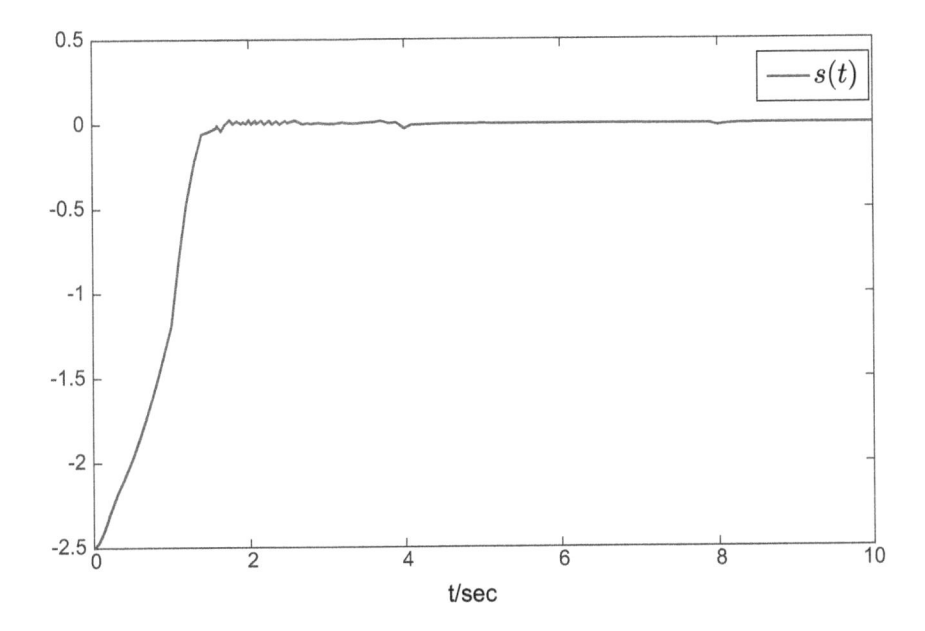

FIGURE 7.4 The curve for switching surface $s(t)$.

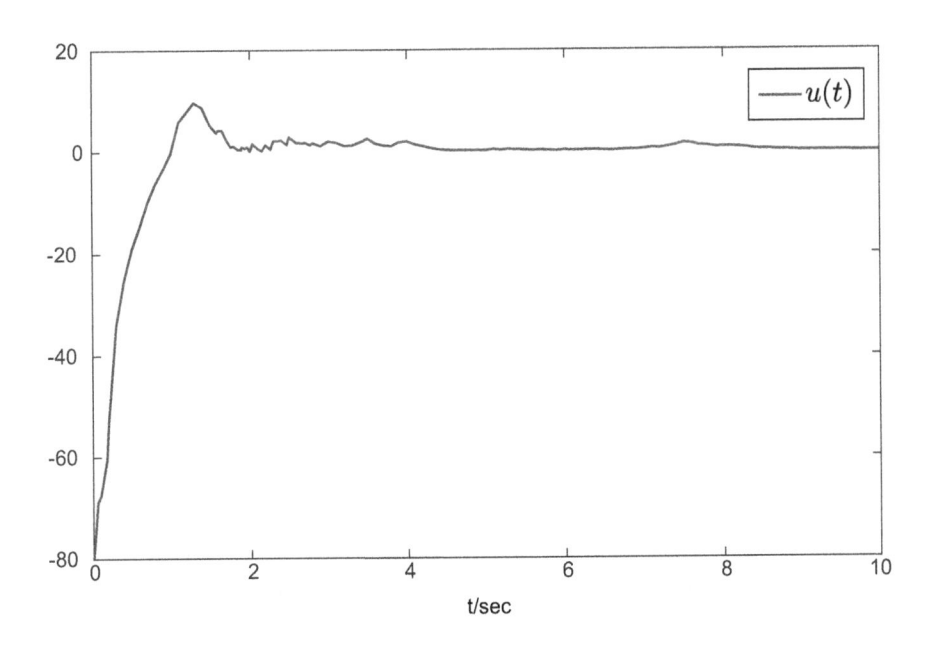

FIGURE 7.5 Control input $u(t)$.

Example 7.2

Recently, the power converter is widely investigated in the literature (see for instance [19,20]). Here, let's consider the typical PWM-based DC-DC buck converter provided in Ref. [20]; an average system model is described as:

$$\begin{cases} L\dot{i}_L = \sigma E - v_s, \\ C\dot{v}_s = i_L - \dfrac{v_s}{R}, \end{cases}$$

where i_L, v_s, L, C, R and E are the inductance current, the average output capacitor voltage, the circuit inductance, the circuit capacitor, the circuit load resistance and the voltage of the external source, respectively. $\sigma \in [0,1]$ is the duty ratio function taken as the control signal of PWM. Correspondingly, let $x_1(t) = e(t) = v_s - v_r$, where v_r is the reference output voltage. Then,

$$\dot{x}_1(t) = \dot{e}(t) = \frac{1}{CR}(Ri_L - v_s).$$

Defining $x_2 = Ri_L - v_s$, it holds

$$\dot{x}_2(t) = \frac{R(\sigma E - v_s)}{L} - \frac{1}{CR}(Ri_L - v_s) = \frac{R(\sigma E - v_r)}{L} - \frac{R(v_s - v_r)}{L} - \frac{1}{CR}(Ri_L - v_s).$$

Denoting $x(t) \triangleq [x_1^T(t) \ x_2^T(t)]^T$ and $u(t) = \sigma E - v_r$, it further leads to the following state-space model:

$$\dot{x}(t) = \begin{bmatrix} 0 & \dfrac{1}{CR} \\ -\dfrac{R}{L} & -\dfrac{1}{CR} \end{bmatrix} x(t) + \begin{bmatrix} 0 \\ \dfrac{R}{L} \end{bmatrix} u(t),$$

In the following, we will consider the above dynamics with three stochastic jumping modes and two deterministic switchings with respect to L and C, respectively. The corresponding values are shown in Tables 7.1 and 7.2. In addition,

TABLE 7.1

Stochastic Parameters

Mode i	Parameter M
1	1H
2	2H
3	4H

TABLE 7.2

Deterministic Switching Parameters

Mode i	Parameter M
1	1H
2	2H
3	4H

the input voltage, the desired output voltage and the load resistance are given as $E = 20$ V, $v_r = 10$ V and $R = 10\,\Omega$, respectively. Considering the above model, which is time-delay-free and disturbance-free. However, the conditions proposed in Theorem 7.1 are also valid for checking the stochastic stability of the above system without taking the time delay into consideration. Therefore, we have the following system matrices:

$$A_{1,1} = \begin{bmatrix} 0 & 0.5 \\ -10 & -0.5 \end{bmatrix}, A_{1,2} = \begin{bmatrix} 0 & 0.5 \\ -5 & -0.5 \end{bmatrix}, A_{1,3} = \begin{bmatrix} 0 & 0.5 \\ -2.5 & -0.5 \end{bmatrix},$$

$$A_{1,1} = \begin{bmatrix} 0 & 0.2 \\ -10 & -0.2 \end{bmatrix}, A_{1,2} = \begin{bmatrix} 0 & 0.2 \\ -5 & -0.2 \end{bmatrix}, A_{1,3} = \begin{bmatrix} 0 & 0.2 \\ -2.5 & -0.2 \end{bmatrix},$$

$$B_{1,1} = B_{2,1} = \begin{bmatrix} 0 \\ 10 \end{bmatrix}, B_{1,2} = B_{2,2} = \begin{bmatrix} 0 \\ 5 \end{bmatrix}, B_{1,3} = B_{2,3} = \begin{bmatrix} 0 \\ 2.5 \end{bmatrix}.$$

Selecting $C_1 = C_2 = -1$ as the gains in the switching surface function, the other parameters λ, μ and the TR matrices are chosen as in Example 7.1. Also, we have the following feasible solutions:

$$P_{11} = 1.8684, P_{12} = 1.9814, P_{13} = 2.1141,$$

$$P_{21} = 5.4592, P_{22} = 5.2990, P_{23} = 5.2184$$

$$W_{11}^1 = 14.7778, V_{23}^1 = 3.6639, W_{31}^1 = 3.6455, W_{33}^1 = 3.6236,$$

$$V_{12}^2 = 3.6446, W_{21}^2 = 3.6744, W_{22}^2 = 3.6272, W_{33}^2 = 3.5157.$$

For simulation purposes, the initial condition is set by $x(0) = [-5 \ 5]^T$ and the ADT chosen as $T_a = 2$. In designing the sliding mode controller, the tuning scalar is chosen as $\rho = 0.01$, and the switching signal is replaced by $s(t)/(\|s(t)\|+0.01)$. Then, Figure 7.6 shows the state response of system by the controller (7.42). Figures 7.7, 7.8 and. 7.9 plot the output voltage, the duty ratio and the inductance current, respectively.

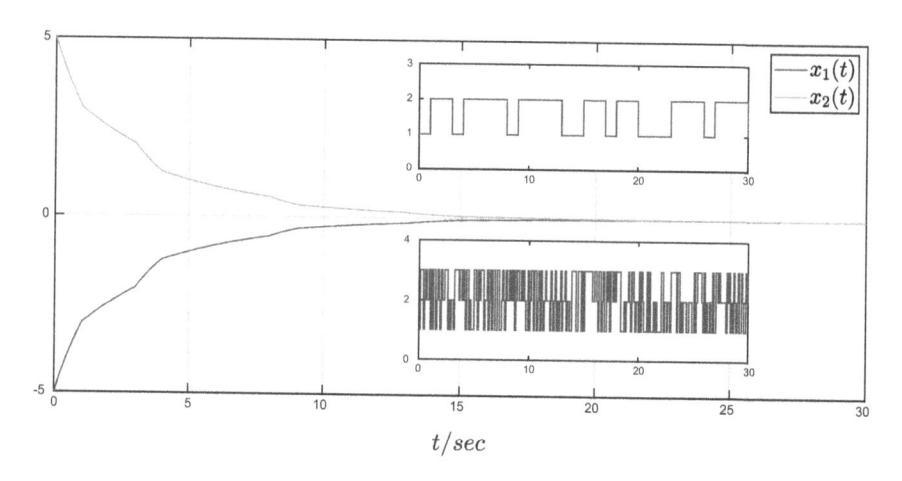

FIGURE 7.6 State response of the closed-loop system.

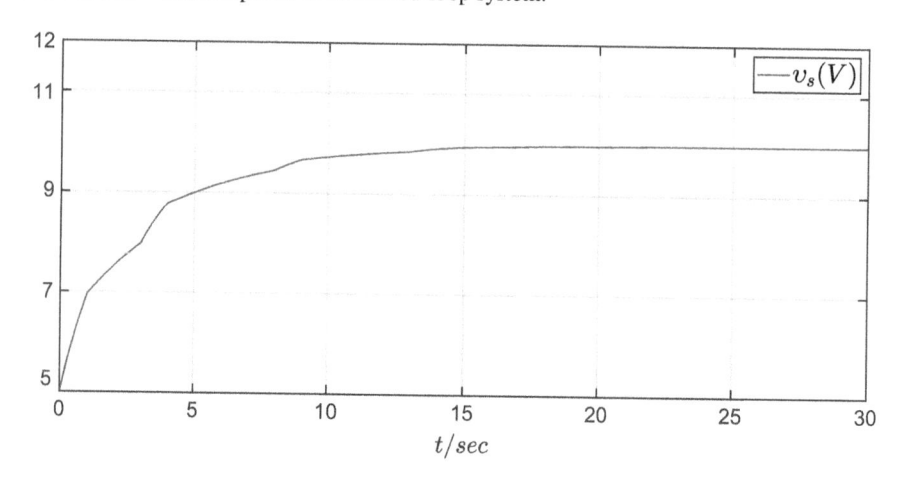

FIGURE 7.7 The output voltage.

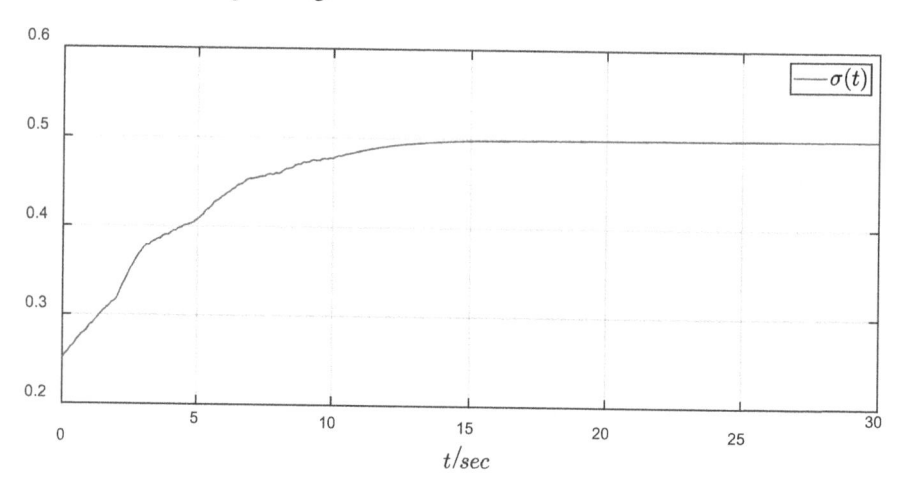

FIGURE 7.8 The duty ratio.

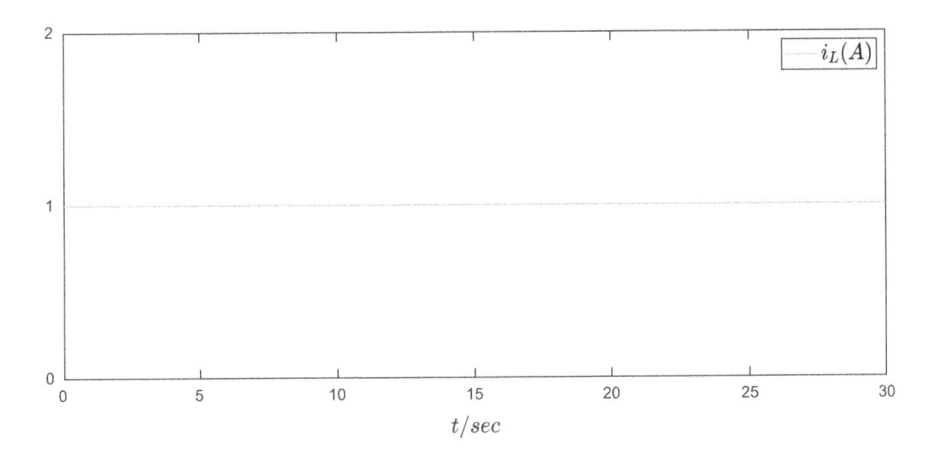

FIGURE 7.9 The inductance current.

7.5 CONCLUSION

Stabilization of nonlinear switching semi-Markovian jump delayed systems with generally uncertain TRs via adaptive sliding mode method has been investigated in this chapter. First, sliding mode dynamics in the low dimensions based on the linear switching surface function has been gained. Then, sufficient LMI conditions for MSES of sliding motion have been proposed. Further, an adaptive SMC law has been developed to guarantee reachability of sliding surface in fixed time. Finally, practical examples have been presented to show the proposed theory numerically. Besides, it is worth pointing out that some issues can be further investigated in the future based on the method proposed in this chapter: for instance, how to deal with the issue of reduced-order SMC design for such models in the presence of mismatched disturbances since the obtained sliding mode dynamics is no longer disturbance rejected.

REFERENCES

[1] J. P. Hespanha and A. S. Morse, Stability of switched systems with average dwell-time, *Proceedings of the 38th IEEE conference on decision and control (Cat. No. 99CH36304). IEEE*, Vol. 3, pp. 2655–2660, 1999.

[2] J. Daafouz, P. Riedinger and C. Iung, Stability analysis and control synthesis for switched systems: A switched Lyapunov function approach, *IEEE Transactions on Automation Control*, Vol. 47, No. 11, pp. 1883–1887, 2002.

[3] X. Zhao, L. Zhang, P. Shi, et al., Stability of switched positive linear systems with average dwell time switching, *Automatica*, Vol. 48, No. 6, pp. 1132–1137, 2012.

[4] X. Zhao, X. Zheng, B. Niu, et al., Adaptive tracking control for a class of uncertain switched nonlinear systems, *Automatica*, Vol. 52, pp. 185–191, 2015.

[5] L. Zhang and P. Shi, Stability, l2-Gain and asynchronous H∞ control of discrete-time switched systems with average dwell time, *IEEE Transactions on Automation Control*, Vol. 54, No. 9, pp. 2192–2199, 2009.

[6] S. Roy and S. Baldi, A simultaneous adaptation law for a class of nonlinearly parametrized switched systems, *IEEE Control Systems Letters*, Vol. 3, No. 3, pp. 487–492, 2019.

[7] X. Liu, X. Su, P. Shi, et al., Fault detection filtering for nonlinear switched systems via event-triggered communication approach, *Automatica*, Vol. 101, pp. 365–376, 2019.

[8] P. Bolzern, P. Colaneri and G. De Nicolao, Markov jump linear systems with switching transition rates: Mean square stability with dwell-time, *Automatica*, Vol. 46, No. 6, pp. 1081–1088, 2010.

[9] P. Bolzern, P. Colaneri and G. De Nicolao, Almost sure stability of markov jump linear systems with deterministic switching, *IEEE Transactions on Automation Control*, Vol. 58, No. 1, pp. 209–214, 2013.

[10] L. Hou, G. Zong, W. Zheng and Y. Wu, Exponential $l_2 - l_\infty$ control for discrete-time switching markov jump linear systems, *Circuits System and Signal Processing*, Vol. 32, No. 6, pp. 2745–2759, 2013.

[11] X. Luan, C. Zhao and F. Liu, Finite-time stabilization of switching markov jump systems with uncertain transition rates, *Circuits System and Signal Processing*, Vol. 34, No. 12, pp. 3741–3756, 2015.

[12] W. Qi, X. Gao and Y. Kao, Passivity and passification for switching markovian jump systems with time-varying delay and generally uncertain transition rates, *IET Control Theory and Application*, Vol. 10, No. 15, pp. 1944–1955, 2016.

[13] J. P. Hespanha and A. S. Morse, Stability of switched systems with average dwell-time, *Decision and Control, 1999. Proceedings of the 38th IEEE Conference*, Vol. 3, pp. 2655–2660, 1999.

[14] V. I. Utkin, *Sliding Modes in Control Optimization*, Berlin: Springer-Verlag, 1992.

[15] E. K. Boukas, *Stochastic Switching Systems: Analysis and Design*, Boston: Springer Science & Business Media, 2007.

[16] J. Huang and Y. Shi, Stochastic stability of semi-markov jump linear systems: An LMI approach, *2011 50th IEEE Conference on Decision and Control and European Control Conference (CDC-ECC)*, pp. 4668–4673, 2011.

[17] E. Shmerling and K. J. Hochberg, Stability of stochastic jump-parameter semi-markov linear systems of differential equations, *An International Journal of Probability and Stochastic Processes*, Vol. 80, No. 6, pp. 513–518, 2008.

[18] X. Zhang and Z. Hou, The first-passage times of phase semi-markov processes, *Statistics & Probability Letters*, Vol. 82, No. 1, pp. 40–48, 2012.

[19] S. Vazquez, J. Rodriguez, M. Rivera, et al., Model predictive control for power converters and drives: Advances and trends, *IEEE Transactions on Industrial Electronics*, Vol. 264, No. 2, pp. 935–947, 2017.

[20] J. Wang, S. Li, J. Yang, et al., Extended state observer-based sliding mode control for PWM-based DC-DC buck power converter systems with mismatched disturbances, *IET Control Theory & Application*, Vol. 9, No. 4, pp. 579–586, 2015.

Outlook

In this book, both the stochastic stability and stabilization problems for a class of semi-Markovian jump systems (S-MJSs) with generally uncertain transition rates (TRs) is first studied by combining the T-S fuzzy model approach and the sliding mode control (SMC) strategy. Then, the proposed theories are mainly applied to single-link manipulator system and RLC circuit system, etc. In the process of control strategy design, an improved sliding surface is proposed, a novel sliding mode observer is presented, and reasonably less conservative criterion conditions are given for the system performance in the presence of more generally uncertain TRs. Similarly, the proposed theories and design schemes can also be applied to some other complex systems, such as neutral systems, singular systems, hybrid system and so on. In addition, this work brings the following novelties in the design of sliding surface and sliding mode controller:

1. For the studied control system, the input matrix is not required to meet traditional constraints, which is independent of fuzzy rules and column full rank, which extends the depth and applicability of the theory proposed in this work to a certain extent.
2. In the scheme of sliding mode observer design, a novel integral sliding surface combined with compensator is put forward. In this way, the nonlinear interference of the system is compensated and the robustness property of the dynamic is enhanced. Besides, it brings convenience for the analysis of various performance indexes of ideal sliding mode dynamic.
3. A type of novel adaptive sliding mode controller is designed. Based on the proposed sliding mode observer, the design of the controller is improved by using the output of the system and the designed observer, which also guarantees the finite-time reachability during the reach phase.
4. For the analysis of the SMC of S-MJSs, this work proposes stochastic stability analysis under the condition more generally uncertain TRs. Compared with previous studies, this paper goes further and lays a solid foundation for future researches.
5. In the first chapter of this work, the stochastic stability of linear S-MJSs with more generally TRs is tentatively presented. Removing the limitation that $I_{i,k}^{\alpha} \neq \varnothing$ and $I_{i,uk}^{\alpha} \neq \varnothing$, the considered TR matrix is extended to more general case than previous studies, which is benefit for further exploration of S-MJSs in the future.

Meanwhile, in terms of SMC of S-MJSs, there are still many problems that need to be further investigated, which are briefly explained as follows:

1. In the implementation of SMC, the selection of each small parameter often depends on the analysis of the dynamic performance and experience of

practical operation, which is hard to achieve a desirable result. Therefore, it is necessary to analyze how to select these tuning parameters, and to analyze how they influence the dynamic response and the mechanism of action. It is an interesting research direction by determining whether they can be selected through well-established optimization algorithms.

2. In this work, the SMC of S-MJSs does not take into account of external noise, such as the white noise, which is widely found in the real-world systems. Therefore, the research in this work can be extended to the study of S-MJSs with the Itô-type stochastic perturbations. The designs of the sliding mode observer and adaptive controller are of great academic value for such systems.

3. Singular characteristics are inevitable under some environmental conditions, and more attention should be paid to the singular characteristics of stochastic systems. For a subsystem j to be activated, the former state $x(t_i^-)$ may not be an admissible initial condition. That is, when the system switches from the subsystem i at point t_i to the next subsystem, the state shows a sudden discontinuous jump behavior. The consequence is that the solution for the whole generalized switching system will not exist, let alone the control optimization of the system. Therefore, it is necessary to analyze the behavior of jumping states in detail so as to find out the mechanism followed in its dynamic process.

4. In this work, the investigation of SMC can be applied to different research directions: for example, the idea of sliding mode observer design can be applied to the analysis and synthesis of Itô-type generalized S-MJSs and networked control system. In addition, the proposed method can be spread to system predictive control, sensor fault reconfiguration and testing, and high-order sliding mode observer design, etc.

Index